Resistance 101:

26 Points Of Contention

Timothy Raymond

This book is dedicated to my lovely cockatoo friend, who, even after decades of captivity, refuses to accept domestication

Keep fighting the good fight.

Table Of Contents

If you're damned if you do, and damned if you don't, then do.

-Guy McPherson

Action is the antidote to despair.

-Edward Abbey

00. Introduction

I have been studying many subjects for a long time. I am approaching middle-age, and I have been socially conscious since I was a child. I've sifted through a great deal of propaganda and half-truths to get to the bottom of the rabbit hole. You are holding in your hands the result of what I have learned from many years of study. Consider this book a starting place—a beginning or supplement to your journey. If one of these subjects interests you, please read the sources I've provided (and anything else you can find), watch documentaries, read books and then share your knowledge with everyone.

Authors such as Derrick Jensen and John Zerzan have already spilled an ocean of ink on modern society's problems. My intention is to not do the same—at least not exactly. I intend to focus on the new breed of apologists advocating "green" solutions, a resource-based economy or anarcho-capitalism, as well as how to deal with people who are either ignorant or in denial about the world around them. I will guide you through these concepts and explain how civilization is affecting you and the world as a whole, and why pre-agriculture living is the only adequate and rational solution.

I decided to write this book because there are few if any anti-civilization reference books available. You can read this book cover to cover or choose chapters bearing subject matter that is important to you. I wanted to make it easy for anyone short on time to pick it up and immediately begin processing the information. I believe that if you can quickly identify with what is written, you can easily relay it to others and then do something about it.

I, like most, am not interested in endlessly arguing academic philosophical points and counter-points. I look for easily understood, common sense answers to problems people are facing today and will face in the future.

In every chapter, I have tried to convey a general, objective overview. Nevertheless, I believe in full disclosure, so you should know that I have taken a stand. I am on the planet's side and I am allied with anyone else who feels the same way. That is the only group with which I can identify.

Although some may classify me as such, I am not an anarcho-primitivist. I called myself that for a long time until I realized that a lot of stigmas and expectations came with that label. Detractors ask: why don't I go live in the woods? Why am I using this computer if I am against technology? I guess those critics have a point if they would rather attack me than acknowledge relevant issues.

Still, many people who point out civilization and industrial capitalism's unsustainability, insanity and cruelty call themselves anarcho-primtiivists, so let us briefly examine the term. First, "primitive":

1: primitive practices or procedures; also : a primitive quality or state
2a : belief in the superiority of a simple way of life close to nature
2b : belief in the superiority of nonindustrial society to that of the present
3: the style of art of primitive peoples or primitive artists[1]

I wholeheartedly embrace a simple, non-industrial nature-centered way of life. Humans lived that way successfully for most of our species' existence. It is not primitive; it is what works. Therefore, definitions 2a and 2b are useful and lack the borderline racism the word "primitive" carries with it.

Calling older, sustainable ways of living "primitive" implies that the foraging tribes living today are misguided for not embracing Western Civilization. I do not believe that type of thinking is accurate, especially considering many people are actively fighting against modernization. There are even organizations set up to promote the idea that civilization is not for everyone. For example, *Survival International*[2] is a group that exists to support indigenous people and hunter-gatherers who are resisting corporate takeovers of their land. I stand with them, and there is nothing "primitive" about not wanting to be absorbed into industrial capitalism.

Concerning anarchy, I only use *anarchy* or *anarcho-* to mean "against modern hierarchies." I personally believe in some rules as long as the current system of power is in place. Laws against rape, torture and murder are useful and should be enforced. However, these laws only have to exist because modern culture enables rape and murder. In other words, we need protection by the system from the system.

Without civilization and the circumstances it creates, no laws would be necessary. "Primitive" societies had unwritten laws—but not police or politicians—since everyone in a local community depended on each other. If someone overshot the land or otherwise threatened the collective's survival, that person would be ostracized or outright removed from the group.

In any case, do these semantics even matter? What good does in-fighting among people who care do when very powerful forces are united against us? People are murdering the planet *right now* while we are tearing each other down. If you care about a sane and sustainable future, we can work together. All that's important is that we educate ourselves, bring like-minded people together, and resist.

Additionally, I will not focus on the degree to which we should scale back technology or science. I do not have the answers as to the exact specifications of a transition to a sustainable way of living. For example, how do we deal with nuclear power plant meltdowns without continuous electricity? I don't know, but what I do know is that continuing on our current path will lead to our deaths. We can choose to acknowledge this and do something about it, or we can continue to walk blindly off the cliff and take the rest of the planet down with us.

Really, focusing on what we call ourselves, satisfactory degrees of technology or worrying about a transition all miss the point of resistance. We need to start doing something, and it needs to happen *today*. The only argument you need is one with the end goal of living in harmony with nature. As a general outline, perform tasks like you plan on living where you are for the next 5,000 years. Discard everything else.

Dismantling agriculture and civilization are huge tasks, and they are scary ones, but they are necessary. Think of what might happen if we do nothing—if we let the world continue on its current, suicidal course. The silver lining is that because there's so much damage that needs undoing, there's also much that we can all do.

An action I suggest that everyone can do today is to disseminate information—talk to people and share this knowledge of resistance. Don't be an alarmist, and don't be afraid. Just tell the truth. It is literally the least we can do for our fellow Earthlings who have suffered so much because of us and our industrial revolution.

I hope that you can find something to relate to in this book—something that speaks to you, your friends, your family, or the land you love. Use that sense of dread, urgency or whatever emotional reaction you have to fuel your motivation to tell everyone you know. Most of them will have never even considered this way of thinking, which means the dominating culture is doing its job. For you see, civilization cannot continue to rape and murder the planet unless most of us comply.

Moving from where we are now to a sustainable future is going to involve some pain, but waiting to do it will involve even more. Do not wait until it is too late. Do something today. It will take all of us doing more than buying energy-efficient light bulbs or driving electric cars to save ourselves and the planet, and the longer we wait, the more suffering we allow and implicitly support.

Join with me and let's make something happen.

Notes:

1) Primitivism Definition:
   http://www.merriam-webster.com/dictionary/primitivism
2) Survival International:
   http://www.survivalinternational.org/

01. Anarchy

I have asked many people if they believe the government has their best interests in mind. I have never received a positive answer. The reason for that is because it doesn't, and the evidence is all around us. Every day, two hundred species go extinct.[1] The climate is changing. A great number of us are suffering. This is despite monolithic governments everywhere claiming the right to manage us and the planet and also claiming that they are doing a good job. It is time for us to consider an alternative that is becoming more obvious by the day: *anarchy*.

Often, when people think of anarchy, they get a little nervous. After all, isn't anarchy a bunch of men dressed in black clothing throwing rocks through windows? Isn't it crowds rioting in the streets, panic, disorder, martial law, rape, murder, looting and total societal collapse?

Well, no; it isn't. Anarchy is simply the absence of hierarchical force. Unfortunately, the definition most people think of is as follows from Merriam-Webster's site:

a situation of confusion and wild behavior in which the people in a country, group, organization, etc., are not controlled by rules or laws[2]

Further down the page, it says, "the absence of government". That is the definition I wish was more popular, and while it is accurate, it needs to be taken further.

As an aside, my only problem with the anarchy movement is that people focus on it so intensely that other issues end up in the background. Anarchy has become a philosophical fetish. In reality, all one has to do is stop believing in government and resist as best as one can without getting legally shot by legal thugs.

Simply, government does not exist. When people think of "government," they should realize that at its roots it is a system of control—a group of people creating laws (opinions with a gun) and a slightly larger group of people executing those laws. Here in the United States, we call those lawmakers Senators and Congressmen; the executors are the military and the police. These titles are all misleading fiction. It is just people in uniforms doing what other people tell them to do.

Think of government employees on one side of a seesaw and everyone else on the other. This is the balance of power. The more people obey government, the higher up the people's end goes where they sit helplessly. If the people knew their power, they would all pile onto the seesaw and send the government flying into space.

It is hard for many to think about a life in which there is no government. We see police, fireman, and a few other "helpful" agencies, plus we are taught in government-funded schools the conclusion that government is right and regular. Personally, having been through preschool, twelve years of public schools and then eight years of college, not once did a professor encourage me to question government's existence and motives. I had to do that on my own. I would be willing to bet your experiences are similar. I would now like to do just that—think about what the government is and what it has to offer.

It is easy to criticize "the government," but rather than being one institution it is really just a bunch of people doing their jobs. They have most likely been taught since birth to obey authority and that the government keeps citizens safe. Most government employees would probably be skeptical and dismissive at the thought of life without laws. They have probably never even considered anarchy. But—think about it. Anarchy is simply a lack of force. That means in almost every instance in your life— from what you choose to eat for breakfast, where you choose to live, who you choose to marry—is anarchy. Every decision you make every day that does not involve coercion or harming others is anarchy.

Now, consider why you pay your taxes. You don't, actually; the language is purposefully misleading. First, you don't "pay" taxes; they are taken from you directly in the form of income garnishments. You must pay property, real estate, and vehicle excise taxes because if you don't, somebody (a "peace officer") will approach you with a gun, kidnap ("arrest") you, and throw in you in a cage ("jail"). If you resist, more and more men with guns will advance on you until you are either immobilized or dead. This—including potentially murdering non-payers—is all perfectly legal and accepted by just about everyone.

Taxation is just another word for theft. Many individuals may see roads, schools and other benefits from government as justification for taxation, but if someone takes something that does not belong to them, it is theft—regardless of how the stolen items are used. A person may think some public services are useful, but what about the tax dollars spent on bombs, bullets and guns for war? What about tax dollars used for bank bailouts, corporate subsidies, and paying pensions for politicians accused of corruption? Somehow, I think you and I could spend our money more effectively on programs that would better serve us and our communities. We might even become friends talking about how to build the roads and educate our children. The point is—we will decide and agree. We will not point guns at each other and use violence.

Even the act of voting for a government official is an act of force. Instead of you and I discussing something and coming to a mutually beneficial decision, we will instead anonymously cast ballots and beg the ruling class to force our preferences into law. Whether or not the politicians will do as they promised (to somebody else) for us is debatable, and what they actually do is even more questionable. Remember, slavery, racism, and blatantly male-favored policies were all once laws. If you need a more modern example, ask a gay person attempting to marry the individual he or she loves. It is illegal in many places, as was interracial marriage not long ago. Moral or not, it is legal, which is outrageous. Why? Because some lawmaker doesn't like the idea, and he has the police to back him up. No one's opinion should stand in the way of someone's victimless happiness.

Besides the passive aggressive nature of voting, no matter who you vote for, the government, and therefore rule by force, wins.

Government, not anarchy, creates lawlessness and chaos. Consider how many people anarchists have killed. (Try to name even one.) Now, think of the blood ocean and body pile from government-sponsored wars. From 1900-2000, the government killed a quarter of a *billion* people.[3] This shows no signs of slowing down, with modern examples including over a million Iraqis killed (many of them civilians) in a never ending terror scare[4], police murdering US citizens on a weekly basis[5], and of course, the overlooked victims—animals slaughtered by the billions due in part to government farm subsidies.[6] I won't include the relentless imperialist destruction of trees, mountains, rivers, the ocean, etc. as it is beyond the scope of this chapter. But, the facts are undeniable: government doing its job is really bad for people and the land they live on. Creatures that get in government's way often end up dead or at the very least (as with animals and American Indians) displaced.

In short, last I checked, no anarchists have nuclear weapons stockpiles.[7]

Now, how many anarchists rape people? (Try to name one.) Probably not many because the government keeps us safe, right? Wrong. More than two hundred thousand men are raped in US prisons every year.[8] The government facilitates this by sub-contracting to private, for-profit prisons as well as jailing the vast majority of "criminals" for non-violent crimes, such as smoking marijuana. One marijuana arrest occurs every forty-two *seconds* in the United States[9], so the opportunities for prison rape are enormous and growing every day—all funded by tax dollars stolen largely from the poor. In abstraction, the ruling class is stealing directly from poor people's families via taxation to lock up members of those families. When do we get to vote on that?

I don't want to overdo the government bashing, so I will demonstrate with one final example: education. Try to think of a world without public schools. Perhaps you feel a little anxiety; after all, how would the children learn their ABCs? Wouldn't there be a literary vacuum? How would businesses form their hiring policies without high school diplomas and university credentials? Eliminating the Department of Education and the entire publicly funded (read: paid for with stolen money) school system seems like a ridiculous and unrealistic proposition.

Well, let us examine the US public education success rates and policies. Brushing aside the archaic teaching model still in practice that colonial settlers would recognize, as with all government programs, education in the United States has been a resounding failure. It is evident in Western countries where IQs have dropped fourteen points in the last century—especially in the United States where tens of millions of "educated" people are functionally illiterate.[10]

This is despite the billions of tax dollars forcibly taken from citizens and spent on schools, tens of thousands of regulatory pages—both state and federal, The Department of Education, No Child Left Behind[11], accreditation protocols, standardized tests, community outreach programs... The list is long and convoluted. The ruling class has had decades to refine and improve academics. Despite ample time, plus the student loan debt recently topping a trillion dollars[12], by all measurable accounts, standards continue to decline. Public education is a bust. Every year, politicians spend more tax dollars yet people's intelligence is not improving.[13]

It is important to note that "primitive" hunter-gatherer tribes were largely egalitarian—that is, they rarely had rule by force.[14,15] Everyone shared in all aspects of life rather than having a designated proletariat. Further, they did not see themselves as masters of the Earth—using nature as one big filling station from which they could endlessly extract. They lived sustainable lives and had deep respect for the land that supported them. This is pure anarchy. After all, can anyone who does not believe in force and hierarchies posit that humans are at the top of the food chain? Force is force; aggression is aggression whether it is human to human or human to anything else. The uncivilized model worked for hundreds of thousands of years and is something humans should look to for guidance rather than eschewing it as archaic.

The bottom line is that any association with government and class hierarchies is one of force and coercion. No productive relationship can ever have this set of values, and constant successful demonstrations in everyday life showcase that people are perfectly capable of making decisions without the threat of violence.

Civilization-born governments cause the very wars, starvation, rape, widespread violence, abuse, cruelty and disorder that they claim to alleviate. Don't you think you and I could talk it out and do a better job?

If you want a great example of anarchy—of a beautiful entity working without government force—spend a little time with your family. Every moment of happiness is one for which you can thank anarchy.

Notes:

1) 200 Species Extinct Every Day:
   http://www.huffingtonpost.com/2010/08/17/un-environment-programme-_n_684562.html
2) Anarchy Definitions:
   http://www.merriam-webster.com/dictionary/anarchy
3) 20th Century Democide:
   http://www.hawaii.edu/powerkills/20TH.HTM
4) Iraq Body Count report: how many died and who was responsible?
   http://www.theguardian.com/news/datablog/2012/jan/03/iraq-body-count-report-data
5) You're Eight Times More Likely to be Killed by a Police Officer than a Terrorist:
   http://www.cato.org/blog/youre-eight-times-more-likely-be-killed-police-officer-terrorist
6) Disturbing Facts On Factory Farming & Food Safety:
   http://www.organicconsumers.org/Toxic/factoryfarm.cfm
7) Anarchy: Frequently Asked Questions by Stefan Molyneux: http://youtu.be/60veZm9ZbyU
8) Prison Rape Widely Ignored By Authorities:
   http://abcnews.go.com/WNT/story?id=131113
9) One Marijuana Arrest Occurs Every 42 Seconds:
   http://www.huffingtonpost.com/2012/10/29/one-marijuana-arrest-occu_n_2041236.html
10) Study: Western IQs dropping:
    http://lesacreduprintemps19.files.wordpress.com/2013/05/were-the-victorians-smarter-than-us.pdf
11) No Child Left Behind, Various:
    http://www2.ed.gov/nclb/landing.jhtml
12) How The $1.2 Trillion College Debt Crisis Is Crippling Students, Parents And The Economy:

http://www.forbes.com/sites/specialfeatures/201 3/08/07/how-the-college-debt-is-crippling-students-parents-and-the-economy/

13) Adult Literacy In America:
http://nces.ed.gov/pubs93/93275.pdf

14) Egalitarian and Resource Conservation In Hunter-Gatherer Societies:
http://weber.ucsd.edu/~carsonvs/papers/41.pdf

15) The Origins of Hunter-Gatherer Egalitarianism:
http://libcom.org/files/Hunter-Gather-er%20Egalitarianism%20by%20Christopher%20B oehm.pdf

02. Birth and Brainwashing

"We do not need books about psychology in order to learn to respect our children. What we need is a total revision of the methods of child rearing and our traditional view about it.

The way we were treated as small children is the way we treat ourselves the rest of our lives: with cruelty or with tenderness and protection. We often impose our most agonizing suffering upon ourselves and, later, on our children."

--Alice Miller[1]

I have heard talk in the mainstream media and among people in their thirties and older about how parents these days simply do not discipline their children properly. If they were spanked and intimidated, these children would not get arrested, get their peers pregnant, never use drugs and get great grades in school. There's nothing wrong with a belt or good-intentioned smack.

The problem with that of course is that it's all nonsense. Says the American Psychological Association:

Many studies have shown that physical punishment — including spanking, hitting and other means of causing pain — can lead to increased aggression, antisocial behavior, physical injury and mental health problems for children. Americans' acceptance of physical punishment has declined since the 1960s, yet surveys show that two-thirds of Americans still approve of parents spanking their kids.[2]

This troubles me. I was a child abuse victim. It was horrific and it took a great deal of effort to unlearn negative and destructive behaviors. My experiences aside, I want to pose the question to anyone who harms defenseless children: how can these child victims grow up to not do the same to their children or to other vulnerable creatures?

It seems a common sense notion that teaching a child that it is okay to direct violence toward the weak would result in that child not learning to respect "lesser" creatures, such as non-human animals or trees. While science has not uncovered any definite causal link between spanking—which is nothing but bullying and intimidation—and animal abuse[3], children tend to mimic their parents' behavior. Generally, fat parents have fat children. Religious parents have religious children. And, if caregivers deal with situations with hostility and aggression, it seems very likely children will do the same. Dominating others in adulthood is a predictable outcome.

Regarding bullying, it is also relevant to examine the ruling class's behavior toward the world. For example, how can a parent in the United States tell their children not to bully others when their government is bullying other countries? The government fights wars, which is using force to make foreign nations submit to US interests. The government allows hydraulic fracturing, which sends the message that it is okay to tear down the environment (animals, trees and rivers be damned) in order to extract oil from the ground. The government sends people to jail routinely for non-violent crimes, where prisoners are often raped and assaulted. The government tortures foreign detainees.[4] The list goes on and on.[6]

Modern society has produced a host of harmful behaviors resulting from improperly raised children. For example, most of us were raised to believe that trees are just things in the way of shopping malls and golf courses. Contractors have every right to cut them down. Nothing important lives there—especially the trees themselves. Just like people living over "our" oil, trees are simply

17

things to be removed in the name of progress and furthering civilization.

The reality of course is that trees are alive. They are not just alive; they are miniature ecosystems in and of themselves, plus they support many creatures symbiotically. This information is common knowledge, but no one seems to care. In modern society, businesses and governments cut down one acre of rain forest every four seconds[7], which causes catastrophic harm and death. We teach our children nice words for this: deforestation, clear cutting, etc. Rather than lying, parents should call it what it is: *murder*. Incredibly brutal murder for profit committed daily with impunity.

Normalizing bloodshed has incredible implications on the values parents are teaching their children—especially those parents who use violence as discipline. If children are raised to view trees and the animals living there as expendable, how will these same children look at mentally ill or handicapped people? What if children raised in this manner end up in positions of power? How will they treat those who are "under" them or below them on the financial and social ladders? It seems obvious: children taught predatory values will likely end up as predators.

A further repulsion one must consider is how improperly taught men will view and treat women. In general, males are physically larger and stronger than females. If male children learn that killing trees and weaker animals is socially acceptable, what will stop them from mistreating women—even without realizing what they are doing? Considering the many thousands of sexual assaults and rapes that happen just to women in the United States every year[7], it is obvious that children urgently need better parenting.

Some may posit that the government's influence on society creates aberrant people, which is correct. The answer, they say, is a society built on non-aggressive capitalism. The free market philosophy, which is trade without coercion, may lessen the killer instinct in damaged adults, but it cannot eliminate it. Abstractly, according to capitalism, every inch of the Earth is for sale. Everything has a price—from the sun's rays to mountains to trees to fish in the sea to humans via labor. Parents may try to tell children to be kind to everyone, but unless children see trees, rocks, rivers and animals as equals and individuals rather than as dollar signs, the violence will continue and get worse as the population grows. Children need to think of the Earth as owning humans rather than the other way around. We are guests on the planet. It is not ours to exploit.

This view—that we are all part of the Earth and equal among it—was how pre-agricultural humans lived. It enabled them to survive and often thrive for hundreds of thousands of years. These societies often drew, for example, US colonial settlers to the American Indians' way of life[8]—far away from a much freer market than we have today.

Uncivilized societies seem more civilized than ones existing today. For example, they did not overfish; instead, they realized that oceans and rivers are alive and should be protected. (Look at a drop of water under a microscope for proof of life.) They did not clear cut forests; they realized that trees and those living in them had beautiful lives of their own. They did not blow off mountain tops with explosives for aluminum and coal. They did not create common household cleaners (read: poisons) for which there is no antidote. They did not overpopulate and throw babies into dumpsters because of their economic status. They did not drive species extinct. They did not drop bombs for oil. Instead, they lived sustainably.

I do not want to romanticize that there was no hardship or violence before agriculture and domestication. That would be preposterous. Instead, I think it would be valuable if people today would study what worked in the past—what *really* worked, rather than encouraging corporal punishment. Children need to be taught to value each tree and animal as an equal and as an individual. Each human, each rock, each ant—all deserve respect as creatures created out of the same stuff. I believe these values, taught early, will go a long way in drastically reducing violence in society and improving the general population's mental health as we move closer to living as nature intended.

If you have kids, make it happen.

Notes:

1) Alice Miller:
   http://www.alice-miller.com/interviews_en.php?page=6
2) The case against spanking:
   http://www.apa.org/monitor/2012/04/spanking.aspx
3) Animal Abuse:
   https://www.ncjrs.gov/html/ojjdp/jjbul2001_9_2/page7.html
4) Childhood Bullying by Stefan Molyneux:
   https://www.youtube.com/watch?v=_KxUkRdjD3k
5) Bullying: http://www.stopbullying.gov/
6) An acre of rainforest lost every four seconds:
   http://www.birdwatch.co.uk/channel/newsitem.asp?c=11&cate=_11061
7) Rape Crisis Center Statistics:
   http://www.rccmsc.org/resources/get-the-facts.aspx
8) Virginia's Early Relations To Native Americans:
   http://www.loc.gov/teachers/classroommaterials/presentationsandactivities/presentations/timeline/colonial/indians/

03. Capitalism

The economic system in place today is not capitalism. It is fascism, which is the merger of state and corporate power. The ideal capitalism definition, according to Merriam Webster:

an economic system characterized by private or corporate ownership of capital goods, by investments that are determined by private decision, and by prices, production, and the distribution of goods that are determined mainly by competition in a free market[1]

The immediate flaw in this definition is "corporate." Corporations are legal entities—fiction that the ruling class invented to absolve themselves from responsibility. After all, if a corporation break the law, it is difficult to blame it on one person, which means crimes go unpunished. I see this often on the news especially concerning banks and big businesses causing blatant environmental damage.

The second problem is that this definition leaves out the word "profit", which is vitally important. In capitalism, without profit, there is no incentive. Profit drives the capitalist system, and without it, the entire thing would quickly self-destruct.

Regardless of these two flaws, let's assume that tomorrow we could put unregulated capitalism in place. One of the hallmarks of "free-market capitalism" is the non-aggression principle, which means nobody forces humans to buy, sell or trade. I trade you a dollar for a widget, and we are both happy. I want your widget more than I want my dollar and vice versa. Everyone wins. However, capitalism's entire purpose is to generate profit, to make money, or any other number of euphemisms that translate into "make living things dead." Regardless of philosophy, I think that if a business, in pursuit of profit, cuts down a forest, that is aggressive against the trees, the animals, and the land. The Earth is the third party in a "voluntary" trade situation, and capitalism has been and continues to be quite aggressive toward it—going so far as to ignore a physical reference in the fictional monetary world.

In addition, one's very survival in a capitalist system depends on creating wealth for oneself. In other words, we need to make money to live on this planet, the "cost of living," which is an actual economic metric. (Stop and think about that for a second. What kind of culture are we to consider that normal and justifiable?) Common sense would dictate that if a person needs to be aggressive or show force to make money in order to feed his or her family, that person will discard the non-aggression principle to meet his own needs. When a person is backed into a corner and is facing market correction, it seems likely that no moral justification against force would stop that person from doing whatever it takes to feed his or her children.

Paraphrasing author Derrick Jensen, capitalism is nothing more than making living things dead. This process has many euphemisms like "making money", "resource allocation", "expanding business", and so on. Anarcho-capitalists (ancaps) are quick to point out that making living things dead is okay so long as no human is forced to do anything. If we rational people do not like that system, we can just go live in the woods, they say. We do not have to participate. Well, let us examine this briefly.

First, if I have to go live in the woods to escape that system, capitalism is coercing me to leave. What if I am okay where I am already living? Are my choices to escape or be a capitalist? Seems forceful.

Second, since capitalism's entire world view is that the Earth is a resource and every inch of the planet is for sale, how would I go about moving into the woods without violating somebody's property rights?

I read an article about one couple who actually tried to move into the woods and live off the grid. They built their house to be fully self-sufficient, and, after locals found out, the "authorities" are telling the couple that they must demolish their home.[2]

Now, I am fully aware that "the state" is the one claiming the home owners are in violation of some sort of legal edict, plus in a free system, the state would not be there to force anyone to do anything. However, this does not stop ancaps from continuing to tell me to go live in the woods under the system in place today.

Third, and this is the most obvious of the points, non-humans already live in the woods. Humans, no matter how "free-market", have created lines dividing where they live and where plants and non-human animals should live. In order for me to relocate to the forest, I would have to encroach on whoever is living there. Of course, apparently since I would not be affecting another person by doing this, displacing trees, birds, bugs and bears is fine and not forceful.

In addition, despite any philosophical justifications disregarding non-humans, I will state that ideal, free-market, for-profit capitalism *is* forceful to other humans. For example, if someone dams a river in order to generate and sell electricity, that dam affects humans living in the now flooded plane behind the structure and humans living downstream of the now dried up river.

If capitalists deforest an area or overfish the ocean for profit, it seems a good assumption that anyone dependent on that land or fish will be affected—forcefully, if you will. Let's take it further. If the people on that land who had been living there for generations were considering having children but the capitalists drove those people away, would that be considered forceful to the unborn?

24

If capitalism threatens a species, as long as it is not humans, according to my understanding it is not aggressive to other humans. But, what if that species is bees? What if those bees pollinate plants? What if a capitalist company put chemicals on its seeds that are at least partially responsible for bee colony destruction?[3] Well, capitalists do not consider that forceful since, you know, it's only bees. But without bees pollinating plants, plants do not grow, and many humans will starve to death. Starvation and death seem aggressive to me.

If other capitalist companies blow up mountain tops, drill into the Earth for oil and minerals, and limitlessly take without giving, nobody knows how far reaching that could possibly be to other humans. The Earth is made up of cyclical systems. The planet recycles everything, which means actions on anything affect everything. Or, we can use the archaic "food chain" model to say that if capitalism breaks link after link on a chain, it is eventually going to affect other humans. Am I the only one who thinks that is aggressive toward other humans?

In one way or another, we are all connected via the natural world, not the fictional monetary world on which capitalism runs. You do realize that money is fictional, right? Somebody made it up. You can go ahead and create millions and millions of dollars, yen, or pesos, but it is just make believe. The only real reference that matters in any way is one modeled on the real world.

Finally, the absolute worst aspect of capitalism is the utter hubris and arrogance of the system at its core. Resource management and distribution? We humans really do think a lot of ourselves to actually entertain the idea that we not only should control the planet, but that we have the right to do so. We use so many justifications, such as science, to proclaim ourselves master of the Earth. But, the reality is that the Earth owns us. We will all die and be recycled into something new. It has happened to every other person who thought they were in control.

Try as they might, humans will never mold the Earth to be suitable for Western Civilization. The Earth will fight back forever, and since despite what most of us think, we are not separate from it, the Earth will eventually win this futile war humans have declared on nature.

Given enough time, the Earth can and will recover from anything—even us. Only sustainable, Stone Age-level living might save us.

Notes:

1) Capitalism Definition:
   http://www.merriam-
   webster.com/dictionary/capitalism
2) Eco-couple told to pull down their 'hobbit home'
   made entirely out of natural materials . . . but
   without planning permission:
   http://www.dailymail.co.uk/news/article-
   2382684/Charlie-Hague-Megan-Williams-told-
   pull-hobbit-home-entirely-natural-
   materials.html#ixzz2tOn3n3Eu
3) What Really Happened At Monsanto's Bee Confer-
   ence: http://www.fastcoexist.com/1682672/what-
   really-happened-at-monsantos-bee-conference

04. Domination
(The Non-Aggression Principle)

One of the cornerstones of libertarian and free-market capitalist economic theory is the non-aggression principle (NAP) or non-initiation of force, which prevents harm and protects property rights. Here is a common definition:

"aggression against the person or property of others is always wrong, where aggression is defined narrowly in terms of the use or threat of physical violence."[1]

People who support this theory claim that it will solve today's problems: everything from eliminating governments to initiating widespread peace. It sounds nice; however, huge irreconcilable flaws make the NAP invalid and unusable. Let me explain.

First, the theory does not have a full rationalization for not harming animals.[2] That is, the only definite creatures NAP protects are humans. The answer, according to some NAP supporters, is the non-sadistic principle (NSP). This is how humans should handle non-human animals and the rest of the world: don't torture them.[3] The problem is—what is torture? What is the lower and upper limit? Does "non-sadistic" mean that is okay to murder and displace animals so long as the brutality is swift? I haven't been able to find a suitable answer. NSP shows the ugly underlying truth to capitalism: be human or get out of the way.

Second, what guarantee is there that anyone will follow the non-aggression principle? With a hope and a prayer, free market capitalists claim that without government coercion, individuals can do anything they want so long as it does not harm another person.[4] What exactly is "harm" and how will the principle be enforced? This is anyone's guess.

NAP adherence relies on pure speculation. Society would be free from coercion, so there would be no government police or courts. Some say dispute resolution organizations (DROs) will be monetarily incentivized to administer justice and, needing to stay in business based on reputation, DROs would eschew corruption or favoritism.5 To me, this seems far-fetched. Once money and power are involved, organizations seem to find ways around bad press and shoddy work.

The answer to bad press and shoddy work? Better education and child rearing. Again, this sounds feasible, but what exactly is better education? What standards would one follow? Would there be specific classes on monitoring DROs? Different cultures have wildly varied definitions of what is wrong or right, which further complicates the issue. Moreover, one need only briefly consider the thousands of ways insurance companies, businesses and courts today bend the rules to realize that faith in DROs may be improbable. With enough currency and influence, rules tend to become optional. Rich people generally don't like becoming poor.

Before suggesting society implement DROs, I think capitalists need to answer a few more questions. Will DROs imprison people? If so, will these prisons be for profit? Will there be an appeals DRO? By what standard will people be judged? Is it if one harms another? What is "harm" and what about cross-culture interpretation? If one person steals from another but causes no physical harm, will the DRO hear the case? Again, anarcho-capitalists offer nothing but conjecture. Personally, I find the profit-driven security/enforcement agency model unsettling. (Type "Blackwater Scandals" into a search engine and you will see why.) I have little faith in protection and justice via contracts, which are just pieces of paper—especially when there is money to be made.

Third, I take great issue with any human's claim to property rights. The Earth was here long before people evolved. What right do we have to claim ownership of anything? In most cases, humans must fully control a piece of land before claiming it as theirs. This most likely involves driving unwanted non-human animals and plants off of it. Correct me if I'm wrong, but if anyone can claim ownership of the land, it is those creatures driven off. In addition, can't the land own itself? Soil teams with life. Owning it means enslaving it. In short, "property rights" is another invented euphemism to excuse and promote domination. It is human-made delusion that is unjust and bigoted.

Really—don't people understand that "rights" are just made up by humans to serve humans? They are fiction and should be treated as such. The only standard by which humans should live is how it will affect those around them. Seeing property through the victim's eyes makes displacing flora and fauna much more difficult—as it should be.

Lastly, the largest concern by far with NAP is that it favors humans. As discussed above, handling non-human animals under free market capitalism is anyone's guess. NAP does not explicitly hypothesize a plan to not only to protect the natural world, but also to respect it. It does not acknowledge that trees, rivers and dirt are alive and therefore should not be harmed. It does not prohibit humans from building cities through living forests, mining metals and disturbing animal habitats, manipulating the ocean and destroying marine life, or staying out of the sky to defend birds. It would still allow turbines, dams, toxic emissions, non-human enslavement (so long as it isn't torture) and ruthless domination of nature. Capitalists may assure us that things would be much less costly in a free (for humans?) society, so harm to other earthlings would be less than today, but I find that unacceptable. Any plan for the future must include nurturing and protecting the planet and everything on it.

I propose an alternate arrangement: ditch the money and study the egalitarian societies in which our ancestors lived for hundreds of thousands of years. Since no profit motive existed, hunter-gatherers needed no impetus to protect their lands. They instead entered in a relationship with flora and fauna, realizing they were small parts of a larger whole. Justice in hunter-gatherer tribes may have involved shunning greedy individuals or banishing them outright.[6] Everyone needed to eat, and therefore everyone worked together to ensure a healthy land base. Therefore, in all probability, needing to police anyone was rare. (They certainly did not have for-profit prisons to fill.) This is not theoretical considering how long primitive peoples thrived—unlike any free market capitalist ideals,

Anyone engaging a capitalist in NAP's inconsistencies will probably need to defend his or her lifestyle. "Why are you using the Internet?" "Aren't you harming animals by using electricity?" Unfortunately, I struggle daily with the guilt of being a human living in civilization. From toilet paper to modern food, my existence damages other creatures. (I hear this often from anti-civilization critics; try debating without being called a hypocrite.) I take comfort in two things only: that I did not create this death culture and that I am doing something about it. Withdrawing from it may make someone feel good, but it makes spreading this message of caring, compassion and resistance very difficult. We need instead to use their words and resources against them.

Although learning about, advocating and applying resistance to the dominate culture is heartbreaking work, I highly encourage activists to continue teaching others this vitally important information. A lot of bears, salmon, pine trees and rivers are depending on you.

Even if the chance of bringing this culture down is a fraction of a fraction of a percent, isn't it worth it to try?

Notes:

1) NAP Definition: http://mises.org/daily/3660
2) A Libertarian Replies to Tibor Machan"s "Why Animal Rights Don't Exist":
   http://www.strike-the-root.com/4/graham/graham1.html
3) The Looming Intergenerational War:
   http://youtu.be/WgO89b3ZoiE
4) The Non-Aggression Axiom of Libertarianism:
   http://archive.lewrockwell.com/block/block26.html
5) The Stateless Society:
   http://archive.lewrockwell.com/orig6/molyneux1.html
6) Key Issues in Hunter-Gatherer Research:
   http://books.google.com/books?id=wbsIXAMBIW4C&pg=PA428&lpg=PA428&dq=hunter+gatherer+justice&source=bl&ots=PTIailBNl4&sig=cPCfkq4qoUBVLRXaNNHsgqW6yhA&hl=en&sa=X&ei=-UidUpvYIofMsQSe-lYL4Ag&ved=0CEgQ6AEwBDgK#v=onepage&q=hunter%20gatherer%20justice&f=false

05. Electricity

When I discuss simple living and modern reform with people, the most often raised concern is how life could not possibly continue without electricity or other energies. After all, the entire modern Western society relies on cheap, readily available fossil fuels. (Just think of the panic in your area when the lights go out.)

Before addressing this apprehension, realize first that electricity truly does keep us alive today. Electric impulses regulate our heartbeats, move through the body to stimulate muscles, and are crucial in brain function. In this capacity, power is vital.

On the other hand, electricity to run coffee makers, vivisection labs and adding machines is entirely unnecessary. It is because of capitalism that these things and the need to power them even exists. If humans suddenly stopped selling life for profit, consider the vast amount of buildings that would not need to be lit and machines that would not need to be run.

As an example, let us consider what it takes to power an electric chair—one used to kill humans for violent offenses committed in reaction to a violent society. (Could any pre-agricultural people possibly fathom this device or the need for it? Only a civilized mind could have thought it up.)

How does one power the murder device? Simply, a utility company needs to extract fossil fuels from the Earth to either generate the electricity directly or to power machines that smelt metals or otherwise manipulate materials for alternative energies as with solar panels. Getting fossil fuels from there to here can be achieved through hydraulic fracturing, which requires clearcuts and massive water and air pollution, or extraction with huge machines and then transporting the oil or gas thousands of miles—all the while spewing toxins into the air.[1] There is of course the inevitability of oil spills and wars over who owns which resources, but that is beyond our purposes here.

As an aside, once a process involves metals, one has to consider mining. Miners—especially in third world countries—must constantly fear methane gas explosions, cave ins, lung disease from dust, and of course, companies cutting corners on equipment and safety to save a few bucks. As with fossil fuel extraction, mining has a bloody history of killing and exploiting humans and non-humans alike.[2]

In order for a company to deliver electricity to its customer, which in this case is the prison containing the electric chair, it helps to have a government and a military. Both can use edicts and aggression to take miles of land from indigenous peoples or flora and fauna in order to set up infrastructure. The most common delivery methods today are utility poles, which are made of wood (dead trees) and power lines, which are made of metals (see above). Since infrastructure encourages population growth, humans will most likely begin destroying more land in order to make room for building more houses. Of course, this requires a great deal of dead trees, forcibly extracted metal, and concrete, which is stone and other materials, plus lots of fresh water, mixed together using electricity.[3] It is very likely that the rocks and water in concrete used to be some non-human's home, as with the trees and mine sites.

Now, a company has two cost-efficient options: use fossil fuel electricity to generate power down lines and into dwellings, or dam a river and use the water to do the same. In the first instance, coal makes steam, which activates turbines, which generates electricity. As for hydro power, the water behind the dam (the reservoir) becomes pressurized, which turns the turbine blades.

While "clean coal" pollutes the air, the dam scenario is actually far more destructive to the environment. First of course is the massive amount of materials needed to make the dam—largely concrete and metals. Next, dams disturb travel patterns of whatever fish live in the water. The reservoir needed to build pressure behind the dam kills anything that lives there—the bodies of which decompose and release gasses that contribute to climate change. Those dead bodies also pollute the water, which degrades its quality and negatively affects anyone downstream. In short, dams decimate the local environment as well as anyone dependent on the rivers dams block.[4,5] Would you drink from or bathe in water something died in?

The last step in the electric cycle is collecting the power at a home or business. In the prison that houses our electric chair example, many miles of infrastructure—mostly metal-based wires—conduct the electricity from the utility company to the chair. (Don't forget the massive concrete and metals needed to create the actual jail.)

In short, generating electricity may benefit humans in some ways, but it creates a huge amount of collateral damage—usually dead bodies. A transition to a sustainable way of living that does not involve electricity would be incredibly healthful and beneficial to the planet as a whole. To hell with our modern society. It is built on blood. I think we can live without our smart phone addiction if it will save a few billion lives—including many humans in third world countries suddenly freed from war and industrial capitalism. (Remember, it takes a lot of electricity to kill a lot of people.)

Naturally, some people may point out solar, geothermal, wind, tidal or other types of electricity. Regardless of how comparably "clean" these alternative energies are, in order to put the infrastructure in place to generate power, humans will still need to figure out how to eliminate the body count and colossal environmental devastation. Solar panels, for example, require a substantial amount of copper wire, which of course, has to be mined. Windmills require plastic, which comes from oil—

a fossil fuel. Geo-thermal requires massive drilling projects. Tidal means more dams and dead marine life. (I could discuss nuclear power, but it is easy to dismiss that as the insane, incredibly myopic thing that it is. Creating poisons with half lives of thousands of years to facilitate watching TV is not worth discussing, plus it does the same routine environmental damage as the other energy sources[6].)

Pre-agricultural people accepted what the Earth gave rather than assuming they had a right to manipulate atoms for their selfish purposes. Electricity, when viewed in its true light, is just control. How many people would be freed from computer-based jobs without it? How many sweat shop factory slaves would exist without cheap energy? How many electric chairs, nuclear missiles, crack houses, child sex worker dens or factory farms could exist in a society without electricity? None. Relieving ourselves of electricity would be a good first to restoring sanity among ourselves as well as halting modern culture's systematic destruction of the planet for critically inane reasons.

Someone, somewhere is saying, "But.. Without electricity, you will die." Maybe, but my one life is not important in the grand scheme of things. I fully admit—I have no idea how to live without electricity. I am saving as much as I can to attend survival school to learn "primitive" skills as soon as possible. I want to learn to live off the grid, and I will do so as soon as capitalism allows. At this very moment, I, like billions of others, have to pay to live—pay for a system that makes me dependent and exploits me.

Regardless of what I am doing, ask yourself: is a television show or lights at night worth someone's life? How about many someones? If you had to explicitly watch corporations wreck the Earth and kill whoever got in the way, would you still champion electricity? Please think long and hard about this and share it with others.

Notes:

1) How Electricity Works:
   http://www.howstuffworks.com/electricity.htm
2) The Dangers Of Mining:
   http://www.pbs.org/wgbh/amex/1900/filmmore/
   reference/interview/miller_dangersofmining.html
3) What Is Concrete?
   http://matse1.matse.illinois.edu/concrete/ware.ht
   ml
4) Hydroelectric Power: How It Works:
   http://ga.water.usgs.gov/edu/hyhowworks.html
5) Dam Problems—Environmental Impacts:
   http://wwf.panda.org/what_we_do/footprint/wat
   er/dams_initiative/problems/environmental/
6) ABC's Of Nuclear Science:
   http://wwf.panda.org/what_we_do/footprint/wat
   er/dams_initiative/problems/environmental/

06. Food

Have you ever been apple picking? If not, you should go some day. It's a wonderful experience. You pay somebody who owns an orchard and they will let you take as many apples home as you can carry. Since the orchard is generally not an industrial operation, you usually end up with organic pesticide-free food.

Recently, a group in Washington State decided to start an "edible forest," which would be open to everyone—sort of like apple picking but at no cost.[1] I appreciate this idea, but while people tend to view it in a positive light, they do not consider *why* such a thing needs to exist. Why do humans need to manipulate the landscape in such a way while the United State throws out ninety *billion* pounds of food annually?[2] Is the answer, especially in light of the obvious obesity epidemic, really more food?

One of civilization's proudest achievements is agriculture, which, in today's world, is clear cutting a piece of land, chasing off or killing the creatures there and then burning whatever is left—right down to the soil. (The soil is alive by the way; it is bacteria-rich.) Next, humans plow the Earth, plant whatever crop is most profitable, and then spend much time and energy defending the land and foreign crops against animals. This can be through spraying chemicals to kill "weeds" and "pests," as well as trapping and/or shooting various herbivores such as deer and rabbits.[3]

Why did hunter-gatherers decide to settle down and domesticate themselves and their surroundings? Many theories exist:

'Few topics in prehistory', noted Hayden (1990) 'have engendered as much discussion and resulted in so few satisfying answers as the attempt to explain why hunter/gatherers began to cultivate plants and raise an-

imals. Climatic change, population pressure, sedentism, resource concentration from desertification, girls' hormones, land ownership, geniuses, rituals, scheduling conflicts, random genetic kicks, natural selection, broad spectrum adaptation and multicausal retreats from explanation have all been proffered to explain domestication. All have major flaws ... the data do not accord well with any one of these models.'

Recent discoveries of potentially psychoactive substances in certain agricultural products -- cereals and milk -- suggest an additional perspective on the adoption of agriculture and the behavioral changes ('civilization') that followed it. [A new theory presents] evidence for the drug-like properties of these foods, and then show how they can help to solve the biological puzzle just described.[4]

Drug-like properties? Well, it feels great to eat cheese because of a chemical called "casomorphin," which, as theory posits, stimulates opiate receptors in humans.[5] Who doesn't like cookies or other sugar-laced substances? A big nap after a huge meal? Count me in. And on top of all this, who wants to struggle to find food or deal with seasonal food shortages? Maybe some of the hunter-gatherers were on to something when they chose dominion over the Earth. Whatever their intentions, the results have been disastrous.

During debates I am often told that I wish for people to starve—that living a sustainable, non-agriculture-based life would mean an end to full stomachs and good health. I believe these objectors are genuinely concerned with other humans, but I also believe that this concern is based on emotion rather than fact.

Today, almost a billion people are starving to death all over the world despite an abundance of food produced via agriculture.[6] One reason for this is because governments offer huge subsidies for farmers to make food, which drives down the price, but much of the food ends up in the trash instead of eaten.[7] In short, it happens for money.

Another hidden reason behind malnourishment is that "food" produced using modern farming techniques is devoid of nutrients. This means that even though people are eating a lot more, the food amounts to empty calories. The intestines absorb and digest very few vitamins and minerals.[8] The soil, which we all depend on for life, simply cannot handle the increased demand. Humans then thought to mix petroleum with soil[9], which harms the environment when it reaches waterways. The chemicals kill the local algae, which causes them to decompose and release oxygen. Fish and other marine life subsist on a very delicate element balance, and when they encounter more oxygen, it results in a dead zone.[10] Also, really—who wants to eat gasoline?

I only want to touch on it here, but has anyone also considered how domestication has affected plants? The floras humans choose as food-worthy don't do well unless we exert considerable effort by feeding and watering them. I believe the plants are trying to tell us something: they do not want to grow where we want them to grow. Additionally, did you ever wonder if perhaps plants get lonely in the isolation we impose on them so we can feed our fat stomachs? Maybe that is why we must so laboriously toil to keep them alive. Who is motivated to live and thrive after being kidnapped and forcibly placed in a foreign land? After all, plants in the wild rely on huge networks of other plants and animals, not segregation, pesticides, greenhouses and sprinkler systems. But, I digress.

Soil quality depletion[11] happened because as humans began exploiting the land to grow more and more food, they also began to create more humans, which required more land, more food and more damage to nature, which lead to more humans, and so on. As forests came down, top soil eroded into rivers or blew away. Many, especially those unwilling to yield their land for growing livestock and agriculture, lose in this arrangement—sometimes up to and including their lives. The only real beneficiaries of this arrangement are some humans in wealthy countries.

Speaking of affluence, another modern-day diet staple is meat. Many first-world citizens eat meat daily or at every meal (even working it into their traditions, such as turkey on Thanksgiving)[12], and countries like China and India, who view the American standard of living as ideal, are increasing meat consumption yearly.[13] To feed all these people, civilization came up with factory farming, which is a horrifically brutal way to "grow" animals. Pen them up by the thousands in tiny, filthy spaces, feed them unhealthful food, fatten them up, and then ship them off in trucks to be boiled, cut up and killed—often while still alive.[14]

One would think that a plant-based diet is the answer, but that is missing the point. It would perhaps do less damage to the Earth, but it still involves monopolizing land and destroying any plant or animal who threatens human food. In addition, when humans began eating grains and other staple crops, their health suffered. From the transition from hunter-gatherer to agriculture, anthropology has uncovered telling evidence of humans being born shorter with their teeth rotting, their bones becoming weaker, and also developing a whole host of new diseases from, among other reasons, communing with animals.[15] This only became worse as time went on, with more humans needing more land and developing hierarchies. It shaped the system that exploits us and the rest of the planet today.

It is probable some hunter-gatherers found themselves starving, and there would certainly be problems if we transitioned too quickly from an unsustainable, agriculture-based way of life to only taking what nature offers. However, considering the current worldwide suffering of human and non-human alike in the form of land base devastation, mass starvation driven by capitalism, and the simple fact that once fossil fuels will eventually run out, it pales in comparison the problems we face if we continue to live this way.

Notes:

1) Beacon Food Forest:
   http://www.beaconfoodforest.org/#
2) Americans Throw Away Ninety Billion Pounds Of
   Food A Year:
   http://www.miamiherald.com/2013/06/17/34560
   50/americans-throw-away-90-billion.html
3) Earth at Risk part 2 Lierre Keith:
    http://youtu.be/wJZrsHoLymg
4) The origins of agriculture: a biological perspective
   and a new hypothesis:
   http://www.ranprieur.com/readings/origins.html
5) Effect of casein and beta-casomorphins on gastro-
   intestinal motility in rats:
   http://www.ncbi.nlm.nih.gov/pubmed/2319342
6) 2013 World Hunger and Poverty Facts and Statistics:
   http://www.worldhunger.org/articles/Learn/worl
   d%20hunger%20facts%202002.htm
7) Farm Subsidy Database: http://farm.ewg.org/
8) Breeding The Nutrition Out Of Our Food:
   http://www.nytimes.com/2013/05/26/opinion/su
   nday/breeding-the-nutrition-out-of-our-
   food.html?_r=0
9) Fertilizer Buying Guide:
   http://environment.nationalgeographic.com/envir
   onment/green-guide/buying-
   guides/fertilizer/environmental-
   im-
   pact/?rptregcta=reg_free_np&rptregcampaign=2
   0131016_rw_membership_r1p_us_se_w#register
10) How Fertilizers Harm Earth More Than Help Your
   Lawn:
   http://www.scientificamerican.com/article.cfm?id
   =how-fertilizers-harm-earth

11) Are Depleted Soils Causing a Reduction in the Mineral Content Of Food Crops?
http://soils.wisc.edu/facstaff/barak/poster_gallery/minneapolis2000a/

12) Meat Consumption Per Capita:
http://youtu.be/Wp51OvHrZNM

13) China and India Driving Global Meat Consumption but West Still Worst Offender:
http://www.ibtimes.co.uk/articles/527128/20131203/global-meat-consumption-increasing-driven-china-india.htm

14) Farm Animal Cruelty:
http://www.aspca.org/fight-cruelty/farm-animal-cruelty

15) A brief review of the archaeological evidence for Paleolithic and Neolithic subsistence:
http://www.nature.com/ejcn/journal/v56/n12/full/1601646a.html

07. Genocide

A common accusation I hear from modern culture addicts is that I advocate mega-death—massive wholesale slaughter of humans. They claim that I am anti-human—a nihilist. They point out that without agriculture, the human population that is hurtling toward ten billion people would not survive. They say that I wish to employ some sort of Nazi-style eugenics program of forced abortions, reeducation and infanticide. And, most egregious of all, I wish to have planet-wide annihilation as my fellow cult members and I simply let all nuclear reactors melt down and sink into the Earth.

The one thing I have noticed with people dealing in accusations and emotional appeals is that they have either run out of logic or they never had an actual argument. Still, as I encounter these wild indictments almost on a daily basis, I will address some individually and the entire notion of anti-humanism as a whole.

First, how can anti-civilization activists be anti-human? They *are* humans. Their mothers, fathers, siblings, children and relatives are human. Their friends are human. They know and interact with many good people. It would seem that someone who is vehemently anti-human would also be a serial killer and suicidal, and I challenge anyone to find a habitually violent person among our anti-civilization fringe of a fringe—let alone a killer.

What people mean with the anti-human charge is that we are anti-modern human lifestyle. This has some accuracy. I cannot speak for everyone, but if anyone believes that human comforts come before nature and innocent human and non-human lives—I am anti-that-particular-human. Nobody's cell phone or jewelry is worth third world children suffering and dying to mine rare-Earth metals.[1] If you ask me, anyone advocating anything that forces people into danger for disposable trinkets is anti-human.

The second round of finger pointing involves agriculture. Ask average modern Western citizens how they would feel about accepting whatever nature gives instead of forcing land to grow a particular crop.[2] Most probably would not even consider that an option. Where would we be without farming? We certainly would not be able to support our colossal (and constantly growing) population. Does that mean I or anyone like me wants to cut off the food supply chain overnight? Of course not. The problem is that eventually, when the fossil fuels run out, the food will soon follow. I absolutely do not want that to happen, which is why I warn people about it. This, again, points to me being a human supporter rather than some snarky curmudgeon.

I charge that anyone who supports civilization is anti-human and supports genocide. Ever heard of world wars? Massive plague outbreaks in close city quarters? Violent crime for money? Numerous child rapes by religious leaders? Mass starvation? School and movie theater shootings? This never happened before our species decided it should own and control the Earth.

And why should we limit our death count to humans? Pro-civilization people apparently favor factory farms, polluting the oceans, climate change, nuclear meltdown, blowing up mountains, enormous dams and deforestation, which probably have racked up trillions of non-human deaths over the last century. Thanks to civilization, quite literally, every life system in the world today is in decline[3], and when pointing the finger at who is to blame, point far away from the anti-civilization crowd. When 200 species per day going extinct[4] ends up at 300 or a thousand, look to modern culture, not me.

All of the genocide accusations, including the Holocaust, starvation, massive death, and nuclear power plants melting into the Earth from neglect have or are happening now. Right now in some cases. Why? Because humans have chosen a suicidal and unsustainable way to live, and they refuse to stop because it might be uncomfortable. For some reason, they attack people like me who suggest that maybe we ought to at least slow down and think about the world we are leaving to our children and grand children. (Damn; there I go being anti-human again.)

Civilization has had thousands of years to prove itself and has, by every measure, failed. It is time we try something else, like going in the complete opposite direction. If civilization recommends it, it is probably wise to do a swift reversal. It may be our only chance of a sustainable future, and it would certainly put the brakes on the incalculable death and misery humans are currently inflicting on the planet and its inhabitants.

I am pro-human. I am pro-you, your family, and your friends. Let's go fix this predicament together.

Notes:

1) Minors As Miners:
   http://www.pbs.org/independentlens/devilsminer
   /miners.html
2) The Vegetarian Myth: Food, Justice and Sustaina-
   bility: http://www.lierrekeith.com/book-ex_the-
   vegetarian-myth.php
3) Data Shows All Of Earth's System Are In Rapid
   Decline: http://www.ipsnews.net/2011/07/data-
   shows-all-of-earths-systems-in-rapid-decline/
4) UN Environment Programme: 200 Species Extinct
   Every Day, Unlike Anything Since Dinosaurs Disappeared 65 Million Years Ago:
   http://www.huffingtonpost.com/2010/08/17/un-
   environment-programme-_n_684562.html

08. Human Domestication

We have enslaved the rest of the animal creation, and
have treated our distant cousins in fur and feathers so
badly that beyond doubt, if they were able to formulate
a religion, they would depict the Devil in human form.
~William Ralph Inge, Outspoken Essays, 1922

I have been living and working with animals my entire life. I've seen them at their best and at their worst. I deeply love them all and most of the best relationships I've ever had have been with non-humans. (Isn't it a little saddening that many humans are so alienated that they keep non-humans as prisoners so they can force friendships? Breeding "companion animals" is big business. My goodness that is depressing)

For decades I thought pets and livestock were normal. Information I uncovered recently broke my heart: they are all domesticated. The "wild" was bred out of them. On purpose, for personal gain, humans have blunted their natural instincts. No animal will voluntarily live in a jail cell—even a very comfortable one like someone's house. Of course, capitalism obliged this human fetish, creating electrified fences, pens, indoor cages, specialized food, training protocols and all other manner of control devices. Some animals respond well to human control; others refuse to be tamed. I have tremendous respect for the latter.

I live full-time with a rescue bird who was taken from the wild illegally because she is very pretty and therefore is an ideal decoration. (The government considers birds property and therefore it is legal to cage them for their entire lives. Few regulations for care or social interaction exist.) I love my bird, but she is very loud. And messy. And fussy. And bossy. But, what can I expect? The Earth was hers to roam--flying to wherever she wanted, completely free, unrestrained. She did as she

49

pleased, and that suited her fine. Her first "owner" soon found out that she would not be tamed. Now, after living with two other people who refused to make the effort to accommodate her, she sits in my living room in a large cage with plenty of toys. She is not a decoration. She is family. I interact with her often, but still, her day is largely spent waiting for me to come home from work. She sees me as an equal—part of her flock. She misses me when I am gone. I hear her call to me when I leave.

Every day, morning and night, I pet and play with her, but even the best situation with me can never be what she once had. She is my prisoner. She can never return to the outside. She would not survive. No rewilding scenario I consider ends well for her. That saddens me greatly. I take solace knowing that she is physically comfortable and that we learn a lot from each other.

Over the years, I have observed this bird doing things that starkly remind me of common human behaviors:

1) Plucking her feathers when she's anxious. This reminds me of humans biting their nails, overeating or doing drugs to deal with anxiety. Wreck the body to quiet the mind. Focus the anguish elsewhere.

2) Repetitive behavior. She will grab a bar on her cage with her beak and run in place in her food bowl. She paces on her perch. She continually screams for me to spend time with her. This is because she's bored and lonely. Humans engage in similar behaviors—especially while at work. We worry, fidget and pace, and then take drugs to make it stop. Even children do these things; intentional bad behavior forces caregivers to pay attention.

3) Not eating or drinking or doing too much of both because she is depressed. Humans mirror this with

anorexia and overeating. It is for the illusion of controlling something—even if it harms the body.

The conclusion is obvious: humans are domesticated too. We lived outside with the rest of the planet for millions of years. We evolved to thrive in nature. Now, most of us cower in offices moving around abstractions wishing away what should be most precious to us: time. We go from phone screen to work screen to TV screen back to phone screen daily in order to dwell in illusion rather than face the control system into which we were born. We destroy our bodies on purpose, fight each other over fictional institutions like politics, and lie to ourselves and everyone else just so we can face another uphill battle tomorrow. We are livestock to the ruling class, who are our tax farmers. We can choose different flavors of coffee but have no say in anything of actual importance— especially whether or not we are free. For most of us, every day is exactly the same. It's a non-stop mental struggle. The animals endure it silently alongside us as we displace, consume and wear them largely for trivial reasons.

We are hopelessly dependent on the system that exploits us. We can't do anything for ourselves. As technology increases, we no longer need maps. We don't have to memorize telephone numbers. We don't need to know how to fix things. We no longer need to parent our children. We don't need to think about food; just go to the store and get it. The television distracts us and the talking heads tell us over and over again that this is the way it should be. Any other way of life is unpatriotic, they say. Civilization is good. Don't worry about where it all comes from or how it will all end.

Still, I think that deep down, everyone knows there is something very, very wrong. What if it all falls apart? What do we do then? How will we survive without all this stuff? We are helpless babes of all ages who think that we know it all, despite the fact that a "primitive" child was for more capable of self-sustenance than any modern adult.

Is it any surprise that seventy percent of Americans are on prescription drugs?[1] In addition, almost eighty million Americans have been diagnosed with a mental illness.[2] Expanding to the rest of the world, psychological problems account for the lion's share of non-fatal diagnoses.[3] Why? Because we are trapped in a civilization cage. It will not allow us to live as nature intended even though that is how we evolved. Our brains tell us something is wrong, but instead of obeying our instincts and solving the problems, we stuff food, drugs and entertainment in the hole where real fulfillment belongs. We go out and spend money on toys to distract us, but in order to get them, we have to work at jobs we hate. The technology keeps us docile and enslaved, and it is a never-ending process.

I hate that this culture has taken away our ability to be self-sufficient. It soaked every meal and small pleasure in blood as our existences in this technological death machine destroys the land we need to live. Those who murder the Earth are rewarded with a high standard of living at the expense of poor people and the natural world. All the while, many who care are marginalized or destroyed.

When this all begins to take its toll emotionally, I go talk to my bird. She helps me through. She makes me keep fighting—even when things seem hopeless. She is a constant reminder of a way of life meant also for humans. She resists domestication just like the hunter-gatherers did. She is a symbol for independence despite having everything humans could imagine she would want except that which is most important: her freedom.

Notes:

1) Study shows 70 percent of Americans take prescription drugs:
http://www.cbsnews.com/news/study-shows-70-percent-of-americans-take-prescription-drugs/
2) The Numbers Count--Mental Disorders In America:
http://www.nimh.nih.gov/health/publications/the-numbers-count-mental-disorders-in-america/index.shtml
3) Mental Health Disorders Now Leading Cause of Non-Fatal Illness Worldwide:
http://articles.mercola.com/sites/articles/archive/2013/09/12/mental-health-disorders.aspx

09. Infanticide

Pro-civilization advocates I have encountered speak often of killing babies. It has become almost a fetish or mantra of theirs. Piles and piles of dead infants would be the result of living without modern medical technology as well as a short harsh existence of people at constant war with each other and nature. If anyone cares for the nascent humans, a return to Stone Age living is simply out of the question. Right? As usual, I must disagree.

The first and most obvious problem with pro-civilization individuals speaking of baby killing is that abortion in today's world is commonplace—so much so that it barely makes the news. Even partial-birth and late trimester abortions—which I believe there can be no question on whether or not the fetus is living. Literally, hundreds of millions of developing humans have been murdered in the last few decades in the United States alone[1]; doctors can even spend entire careers making live babies dead. As population grows (and it does every year[2]), so do abortions. (Note: this is not a pro-life or pro-choice treatise; I am simply pointing out the facts—infanticide is culturally acceptable in the Western world.)

The rest of the civilized planet fares no better—especially the developing countries. Due to socioeconomic pressure, war and poverty, many humans routinely slay their children in dire conditions:

Nearly half of all abortions worldwide are unsafe, and nearly all unsafe abortions (98%) occur in developing countries. In the developing world, 56% of all abortions are unsafe, compared with just 6% in the developed world.[3]

Continuing with the third world, what if the babies die, but the death is slow and painful from starvation? Does it still count as infanticide? For example, of the almost one billion people living with chronic hunger, almost eighty percent of them live in developing countries—especially in Africa, where nearly one in four are undernourished. On top of that, five million child deaths each year can be attributed to malnutrition.[4] That's 13,698 per day, 570 per hour, and nine per second. (Comparatively, under Hitler's Final Solution, between ten and fifteen million people were slaughtered.[5])

In modern-day China, due to the state-enforced one child policy, infanticide—especially among females—is disturbingly common. This includes after birth massacre as well as abortions.[6] Foragers could not have imagined such widespread savagery

Additionally, evidence exists concerning the same practices in India—murdering infants (especially females) due to socioeconomic pressure, poverty and widespread patriarchy, which has lead to female persecution.[7]

Where is the outrage? Where is the public outcry? Where are the concerned voices for the babies filling mass graves in our modern world?

Before moving on from this, I need to clarify that I am not solely focusing on human children unlike my detractors. Humans believe their children deserve more recognition and privileges because they are human, which is circular logic. Obviously, mothers favor their offspring as different and above others, but this is true with all animals. In and of themselves, young members of any species are not extraordinary and ours does not deserve preferential treatment and consideration just because it looks like us.

What's more, why focus only on the young and unborn? At what age does murder become acceptable? If anti-civilization advocates were to include children, adolescents and adults when debating about violent population control, any amount of death in pre-history would not compare to even one century of modern-day statistics.[8] War, poverty, hunger, civilization diseases

55

such as cancer and others take innumerable human and non-human lives every seconds of every day.

I feel the pro-civilization argument focuses too narrowly on humans. Looking elsewhere, worldwide meat production results in billions of dead land animals annually.[9] What about those infants? Standard operating procedure for male chicks is to either grind them up alive or throw them in trash cans to suffocate. The chickens selected for food and eggs are no better off, as they generally live for only one or two torturous years before being brutally murdered for profit.[10]

Further, many *many* more ocean-dwelling babies die each year due to modern civilization—especially trawling. We're talking tens of billions.[11, 12] All of the bloodiest dictators throughout history could have never dreamed of such successful targeted butchery. It is a wonder the oceans are not red with blood. (Maybe it is under all the oil spills and floating plastics.)

Does this genocide not count in the pro-civilization argument simply because these creatures are not human? This type of worldwide bloodbath could have never happened with Stone Age technology.

In order to debunk infanticide among pre-agricultural people, I first had to find information, which was more difficult than it sounds. One would think that rampant murder would disassociate anyone from a future primitive, but it appears to be misinformation and bias. The common stereotype from which the infanticide argument most likely originates from can be found in essays by the seventeenth century statist Thomas Hobbes, who, in reference to anarchic people, wrote "the life of man [is] solitary, poor, nasty, brutish, and short."[13] This is pure propaganda; Hobbes was pro-government and of a common tactic when proposing rule by force is to decry all other societies—especially uncivilized peoples and their independence and decentralization.

Pro-civilization (read: pro-control) advocates point to author Lloyd deMause for confirmation of widespread violence among our Paleolithic ancestors. He presents strong evidence of infanticide, child rape and genital mutilation in antiquity, but large parts of his work focus on post-agriculture states.[14] This actually strengthens the case for dismantling modern society. Domestication causes people to do horrible things to their offspring.

Other anthropologists like Margaret Mead in her works such as *Sex and Temperament in Three Primitive Societies* presented a loving, egalitarian picture of child rearing.[15] Both Mead and deMause have their share of detractors and supporters. I leave it to you to research their theories further. Extrapolating information from dissimilar historians and anthropologists, such as Colin Heywood, Linda Pollock and Larry Milner, one can find just about any viewpoint or agenda concerning infanticide.

Two theories are common: that infanticide uncivilized humans was like abortion to modern people, and that population control and collective survival forced some foragers to execute their newborns.[16] One can even entertain the idea that life was very difficult for some children (despite evidence Mead and others present.)

Regardless, even if all of these criticism of uncivilized life are true to their fullest and most violent extent, the number of ancestral infant lives taken simply cannot compete with the literal billions (possibly trillions) of souls slain under civilization—both human and non-human alike.

Quite simply, the infanticide argument is, at best, contestable, and at worst, half truths designed to demonize an independent, free society, which is in stark contrast to the multiple layers of control throughout civilization. Why would anyone who favors the modern world of "might is right" want to encourage an autonomous, egalitarian culture?

For the final nail in the coffin, one need only look at the extant hunter-gatherer tribes. No data indicating rampant infanticide exists. (If it did, pro-civilization advocates would reference it religiously.) And, while that is not definitive proof, it certainly strengthens the case against a civilized way of life.

Notes:

1) Abortion In The United States:
   http://www.guttmacher.org/media/presskits/abortion-US/graphics.html
2) Population Growth Annual Percentage:
   http://data.worldbank.org/indicator/SP.POP.GROW
3) Unsafe Abortions: Sedgh G et al., *Induced abortion worldwide in 2008: levels and trends*, Lancet, 2012
4) 2013 World Hunger And Poverty Facts And Statistics:
   http://www.worldhunger.org/articles/Learn/world%20hunger%20facts%202002.htm
5) The Holocaust Death Toll:
   http://www.telegraph.co.uk/news/1481975/The-Holocaust-death-toll.html
6) China's Infanticide Epidemic:
   http://www.du.edu/korbel/hrhw/researchdigest/china/InfanticideChina.pdf
7) Female Infanticide:
   http://www2.webster.edu/~woolflm/femaleinfanticide.html
8) Population Clock (includes live death count):
   http://www.worldometers.info/world-population/
9) Animal Production:
   http://www.usda.gov/wps/portal/usda/usdahome?navid=ANIMAL_PRODUCTION
10) Chickens Used For Eggs/Chickens Used For Meat:
   http://www.farmsanctuary.org/learn/factory-farming/chickens/
11) Scientific Facts On Fisheries:
   http://www.greenfacts.org/en/fisheries/#1
12) Marine Fisheries Agency Annual Report And Accounts:

http://www.official-
docu-
ments.gov.uk/document/hc0506/hc15/1526/1526
.pdf
13) Thomas Hobbes, *Leviathan*:
http://oregonstate.edu/instruct/phl302/texts/hob
bes/leviathan-c.html
14) Lloyd deMause, *The Origins Of War In Child
Abuse*:
http://www.psychohistory.com/originsofwar/08_
infanticide.html
15) Margaret Mead, *Sex and Temperament in Three
Primitive Societies* (Various Excerpts)
16) 5 Common Objections To Primitivism and Why
They're Wrong:
http://theanarchistlibrary.org/library/jason-
godesky-5-common-objections-to-primitivism-
and-why-they-re-wrong

10. Junk

Look around you. If you're at your house, do a quick inventory of all the ridiculous items you have collected over the years. Video games, a thousand pieces of clothing, closets full of shoes, DVDs, CDs... Now, how much of this do you need to live? Or even enjoy?

Let's consider a DVD, for example. You go to your local department store, find it in the $5 bin, and for some reason, it feels good to have some poorly-paid clerk bag it for you. You then bring it home and one of two things happens:

1) You forget about it. Admit it. You've left new stuff in your car before. Maybe your spouse will even step on it since you left it on the passenger-side floor.

2) You watch it once, kind of enjoy it and then add it to the ever growing pile of stuff that was once your paycheck.

As far-fetched as it sounds, compulsive shopping is a legitimate modern mental disorder.[1] While some studies may say that addiction origins are largely unknown, I believe I understand them. Although I personally do not have an addictive personality, I have given pause to this subject for many hours. In short, life under civilization is really terrible for a lot of people. Getting high, getting drunk, feeling that rush of buying something or gambling—these are all forms of escapism. It is easy to understand if you've done any substantial suffering.

Consider a single man living in a city. He is coerced into a wage slave job because on Earth, you have to pay to live. His parents were poor, so he has no financial mobility. Therefore, his choices of occupations do not include anything fulfilling. He fights traffic every morning, shows up every day bleary eyed at his dead end job and then goes through the motions—always fighting the urge to run away from the building. He makes few if any connections with his co-workers beyond sports talk and people going on and on about their children. (Nobody really cares about your kids. Trust me.) He then goes home, sits in front of the television, eats poor quality food, and then surfs the Internet for hours hoping and praying to find somebody who or something that can make him feel worthwhile.

After about a decade of doing this—even if the man gets married and has children—he can no longer deny the utter futility of it all. He's bought a car, a decent wardrobe, a next generation video game system, a smart phone, and a bunch of other toys. He's also picked up a drinking habit. Pouring alcohol on his problems helps him tune them out for a while. It also helps shut his head up at night—the constant nagging from his brain that he'd better make something of his life. Of course, advertising tells him every day that he and everything he owns aren't good enough. "Just buy our product and you'll be happy. We *promise*."

After another decade of steady drinking, he develops jaundice. Rather than deter him from alcohol, he keeps up and actually increases the habit. Why? Because life isn't that good. It's overrated. What does he have to look forward to for the next fifty years? Having an infernal noise we call an alarm clock shatter our sleep every day so we can rush to slavery? Getting drunk seems to help; our subject will never give up one of the few things that has given him pleasure after all these years—even if it kills him.

The jaundice begins to progress. It steadily worsens. The doctors find cancer. It is aggressive and terminal. The drinking exponentially increases. The only time anyone sees our subject smile is when he is intoxicated. (Let's assume he's a friendly drunk.) The rest of the time, he still has to show up at his meaningless slave job to pay doctors, bankers, and the government. He has to pretend to care and bottle his rage to function. Still, even with full-time employment, he worries because his money is spent even before he makes it. Drink it away. That's all he can do.

Eventually, the man's wife and kids (let's assume he hasn't battered or completely alienated them) notice he is suddenly much happier and at peace. He's almost completely stopped drinking. Why? Because he's decided he's going to kill himself. He has a time picked out, a place, his last meal... Finally, he really does have control of something. He has rationalized that his family will "get over" his death since he is dying anyway.

After his favorite meal of fried chicken, macaroni and cheese, and a big glass of root beer in a comfortable hotel room, he takes a big handful of the sleep aid pills he's been taking for years. One last drink to go with it. He drifts away, feeling a little bit of satisfaction that the bill collectors and his boss can no longer browbeat him.

Now, really—is this guy a coward? Is he abnormal? He reacted as many domesticated animals do—with self-mutilation and ultimately self-destruction.[2] Why? Because of long-term mental damage and to harness some sort of control. To not feel so impotent and insignificant.

We humans are animals. So, why would we react differently than any other animals to the unnatural situation we are living in? We can lie to ourselves and everyone around us, but we can't fool our biology.

Of course, biology or not, we all seem to have these giant holes to fill that we left behind with natural living. As stated previously, the "answer" for a lot of people is to fill their lives with junk. This fuels the plastic lifestyle manufacturing capitalist machine, which turns living things dead. I guess one could excuse a supposed rational human for destroying himself, but buying all this junk and subsequently dumping it into landfills has very real consequences to the planet.

Plastic, which is made from oil, is not biodegradable and is literally everywhere. I don't think it's possible to even look in a place inside of a building and not see it or one of its polymers. It also litters the landscape everywhere we go.[3]

As an aside, isn't it absurd when somebody tells you not to throw paper in the street because it's littering? What are the street, buildings and cars? Litter. If you're worried about cosmetic waste you may want to expand your definition.

Anyway, humans have thrown away so much plastic that, as one better known example, it created a floating island called the Great Pacific Garbage Patch. This manmade abomination is in the northern Pacific Ocean. Estimates put the size of this thing as high as twice the size of the United States. It gets worse every day.[4] As economies grow, so does trash, which increases the size of this toxic garbage island. The implications of how evil and harmful this is to non-humans and the planet are obvious.

Recently, pictures surfaced of an albatross chick lying dead after ingesting garbage from Midway Atoll, which is near the Garbage Patch. See, after capitalism destroys the land and the oceans, it makes food scarce. Some of our fellow Earthlings will confuse plastic trash for food—in this case, the albatross. They then feed it to their babies. "As a result, tens of thousands of albatross chicks die from starvation, choking, internal bleeding, and poisoning each year."[5] And for what? Burner cell phones? Really nice patio chairs? Those little eggs that contain toys that children buy for a quarter out of machines? Soda bottles? Those red cups college students use for beer pong? Is there any possible way to justify any of this? Are we that barbaric? How many thousands of somebody else's babies have to die for us to satisfy our junk fix?

We really need new phrases added to the discourse. Nothing is ever thrown "away"; it is just relocated to somebody else's backyard where it contributes, at best, to pollution, and at worst, to non-human or poor human disease and death. Harmfully relocated? "Honey, I'm going to harmfully relocate this useless crap I put on my credit card that I'll be paying off for years."

Our ancient ancestors did not treat the land like this because, quite frankly, they were not suicidal. I could go on and on and cite sources, but it's a common sense argument. They lived for millions of years and did not create even a percent of a percent of a percent of the waste civilization has created in just a few hundred years under industrial capitalism.

What's the answer? Is it recycling? Hardly. In order to recycle all of that plastic, it has to be transported to a landfill or collecting station, then transported to a factory, then the factory has to run machines to melt it down and release the byproduct toxins into the air, then the new raw materials have to be transported to another place to shape them, then to a manufacturing facility, then to a warehouse, then to a store and then to somebody's house. Every step of this requires fossil fuels, pollution, and land base degradation.

A bigger problem is that people who recycle feel they are doing well by the planet.[6] It is passing responsibility off to someone else since that is what is hammered into them from birth. I don't have to do anything more; I recycled! Hey neighbor! Don't throw that can away! Toss it into a recycle bin! That's the *green* thing to do! (The word 'green" makes me cringe.) We've done our part to secure the environment for our children!

Keep everything going so long as there are feel-good bins scattered around every community, right? Wrong. The unbelievably obvious solution is to not make this junk in the first place. I have known this since I was a child.

I can remember it clearly. I was six years old and looking around a department store during the Christmas season. I wondered how it was that stores had new plastic everything every single year. Where did it all come from? I asked my parents but they didn't know. My brain signaled that something was very wrong, but I was far too young to figure it out.

I'm not that bright. I promise you. But, if I as a child in public school and the product of government and advertising propaganda could at least consider such an obvious predicament, why can't adults? How is it that nobody wonders why despite futile recycling the US produces 250 million tons of trash every year[7] and how that is insanely unsustainable?

Even if Thomas Hobbes misstatement of uncivilized people as having short, harsh lives were true, it does not excuse the.... I don't even have words for it. Calamitous? Devastating? Earth-shattering? Whatever the adjective, life is now short and brutish for so many non-human Earthlings that have to eat our trash so we can feel better about ourselves for a few minutes. How many more will suffer and die? Tens of thousands of albatrosses every *year*? When will people be concerned—when it's twenty thousand per year? Fifty thousand? When they're extinct? Maybe ten thousand of some other species will do the trick. (Maybe ten thousand humans—as long as they're white men with money.) I highly doubt it, but we can't be naïve enough to think that the albatross is the only non-human suffering as a direct result of our junk habit.

These are not the actions of our ancestors, who wanted and planned to live on their land forever. Whatever their flaws and regardless of their quality of life, they could never have comprehended turning the entire planet into a trash can, or purposefully creating a "design for the dump"[8] economy that destroys everyone's health so we can all play really neat computer games.

Civilization has got to go before, via toxic garbage island, the products we consume ultimately consume us.

Notes:

1) Shopping Spree Or Addiction:
   http://www.webmd.com/mental-
   health/features/shopping-spree-addiction
2) Feather Plucking And Self-Mutilation In Pet Birds:
   http://theexoticvet.com/care-sheets/feather-
   plucking-and-self-mutilation-in-pet-birds/
3) How Plastics Work:
   http://science.howstuffworks.com/plastic.htm
4) The world's rubbish dump: a tip that stretches
   from Hawaii to Japan:
   http://www.independent.co.uk/environment/gree
   n-living/the-worlds-rubbish-dump-a-tip-that-
   stretches-from-hawaii-to-japan-778016.html
5) The Garbage Patch Bird:
   http://www.motherjones.com/blue-
   marble/2010/02/chris-jordan-midway-atoll-
   pacific-trash-gyre-albatross
6) 10 Ways Recycling Hurts The Environment:
   http://listverse.com/2013/01/27/10-ways-
   recycling-hurts-the-environment/
7) Municipal Solid Waste Generation, Recycling, and
   Disposal in the United States:
   http://www.epa.gov/osw/nonhaz/municipal/pubs
   /msw2008rpt.pdf
8) The Story Of Electronics (2010):
   http://youtu.be/sW_7i6T_H78

11. Killers

One of the most successful propaganda and anti-nature organizations that exists is the military. It is quite a staggering feat to convince the general public that certain people, once they dress up in a variety of costumes, are heroes who can do no wrong. Anyone who joins the military has noble intentions, and they definitely are only interested in protecting us and our freedoms, right? I wish that were true. I really do.

(I also wish the ordinary citizens didn't have to swallow lies day after day from the mainstream media about the military and pretty much everything else, but being corporate whores is what newscasters do. If they didn't, perhaps the world wouldn't be in the shape it's in.)

The reality is that many of the people in the first world militaries are kids—teenagers. What a criminal act to allow somebody whose brain is not fully developed to make such a decision! The problem is that potential soldiers have incredible incentives in the form of social acceptance. For example, after the September 11, 2001 World Trade Center attacks in the United States, *eighty percent* of citizens supported war.[1] That's a hell of a lot of peer pressure and not a lot of naysayers. To a kid just out of high school with no idea on how the world works, and with almost everybody's support—plus a healthy dose of government propaganda—the military seems like a good idea. You're a hero once you put on that green costume. But, then you get there. And the reality is nothing like what the news, your friends, and your recruiter told you.

I realize some of this is going to sound harsh, but understand that I tried to get into the military twice. Even though I was rejected both times because of a physical handicap, I too felt the call of duty. I truly feel for all service men and women. We are all victims dealing with psychopath leaders who are willing to spend untold sums convincing people to invade other people's countries.

With that in mind, here are some of my thoughts to help shed some light on the institution that kills everything it touches—the military industrial complex. If we change the language we use, maybe we can save a few lives.

First of all, the mainstream media and everyone else should call anyone in the military what they are: murderers for hire. Even the ones who don't directly destroy human life are exterminating the environment by being involved in the system that generates perpetual conflict to justify its own existence. Their paper pushers, warehouse people, logistics coordinators, drivers and everyone else in uniform are all complicit to the most catastrophic and long-lasting intentional mass murder in history.

It's really simple. If there were no soldiers, there would be no war. If they all simply failed to show, war could not exist.

The military is a huge, bloated operation. In the United States—the most egregious offender—over three quarters of a billion dollars gets spent on "defense" each year.[2]

That entire sentence is a lie. Let me try again. It's going to be long.

A small group of "elected" men and women create laws that allow people in blue costumes to enforce, and IRS agents to steal, nearly a trillion dollars from average citizens to fund the military industrial complex, which is anything but defensive considering the US has hundreds and hundreds of military "installations" around the world in dozens and dozens of countries—at least some of which I don't think are at war.[3] Following trends, I think it's likely that soon the US will spend a trillion dollars on the military in one year—that we know about. Who knows about their "black sites" and off the books operations?

It befuddles me that the average person doesn't think about this. For one moment, forget that this piece of writing exists to try to show you why civilization needs to fall. Let's consider something else. Think about a trillion dollars. That's $1,000,000,000,000. Doing some back of napkin math, if someone had spent a million dollars per day from the year 1 A.D. to today, they still would not have a spent a trillion dollars, not to mention the US debt goes up $41,000 per *second*.[4]

Now, let's assume the government came to its senses (it won't) and decided that itself had to go (never gonna happen), so all government spending stopped, yet they kept collecting taxes to pay back what is owed (which is interesting, since I don't remember signing any contracts with foreign countries or any central banks). Let's assume that everyone decided on a hundred million dollars per day to pay back. That's $100,000,000 each day, one billion every ten days, and about three billion per month. (As a reference, the official media story was that it cost almost $4,000,000,000 to clean up the Exxon Valdez oil spill in the 1990s.[5] What the news meant is that it cost $4 billion to get capitalism up and running again, not that it murdered an Alaskan habitat. Just so we're clear.)

A hundred million per day would pay the US government's $17 trillion debt quickly, right? Well, $100 million per day multiplied by 365 days is $36,500,000,000 or thirty-six billion, five hundred million dollars per year. That's a ton of cash! But—and I blame basic math for this—the problem comes with division. $17 trillion ($17,000,000,000,000) divided by $36,500,000,000 is (rounding down) 465.75 years.

Go ahead. Whip out your calculators. I checked it twice.

The ruling class is laughing at us poor suckers. There is no plan to ever scale back the military or any other government spending. It's only been increasing since the World Trade Center attacks. What does that mean to me, you, and the rest of the sane Earthlings—especially the non-humans?

First, it means that it will be necessary to keep making bombs and bullets. That all requires metal. That all requires mining. That all requires exploitation. That all requires destroying the land base to force poor people into such a miserable job. That all requires a military.

It is really doubtful that indigenous people would willingly forego land that provided for them via clean rivers and air to take a mining or smelting job. Easy enough; just have a war near them or otherwise destroy their land and homes with industrial capitalism. Perhaps stage a skirmish in their backyard and set up a bomb factory near where they live that discharges in the rivers where they fish. That way, the fat white males at the top can doubly destroy the land and its people and everything's gonna be alright according to the news.

Second, because capitalism requires its consumers to keep consuming to keep the whole Ponzi scheme going, and since war is a really great business, there needs to be perpetual conflict. How destructive is war to the environment? Well, wait a second. Just to emphasize, fundamentally speaking, war is a business and a damn good one. It is a natural offshoot of industrial capitalism. I don't know how many times people have told me that the world needs peace before we can move forward (whatever that means). Actually, the world needs to get rid of the incentive to destroy life before there is peace. Industrial capitalism provides that incentive by hiding violence and resource extraction and encouraging people to consume via the media and making it all happen via the military.

Anyway, here's a chilling few sentences from the Peace Pledge Union on what war does to the environment:

A German officer in 1918 described 'dumb, black stumps of shattered trees which still stick up where there used to be villages. Flayed by splinters of bursting shells, they stand like corpses upright. Not a blade of grass anywhere. Just miles of flat, empty, broken and tumbled stone.' The ploughs in Flanders fields still turn up human bones every year.

It gets worse.

[T]wentieth century technology, busily applied to the practice of war, has ensured a more lethal harvest. For example, landmines: planted in millions in war-torn countries across the world, killing and maiming long after wars are over, and denying agricultural use of the land in which they lurk. A Khmer Rouge general called them 'the perfect soldier': cheap, efficient, expendable, never hungry, never needing sleep. But eighty percent of landmine victims are civilians, not soldiers; and nearly a quarter of those are children.

It gets even worse. Please forgive the length of this. I know it's tempting to skip italicized text because the author then has to explain it after, but I assure you, I won't. I won't need to. You need this information with you the next time somebody tells you how heroic your country's military is:

But it's the testing and manufacture of the nuclear bomb which has been responsible for some of the most profound and persistent environmental damage to life on earth. "The complex mixture of contaminants found on many military sites is dynamically moving through the environment,' says a medical expert. Radiation problems affect people near nuclear plants in every country that has them. Repair and maintenance of many installations and equipment are dangerously inadequate.

Nuclear waste is a global problem that won't go away, threatening environmental disaster on a vast scale: its poison, and toxic chemicals which accompany all weapon production, have travelled round the globe in the atmosphere and ocean currents; as well as water and air, they harm earth, plants that grow in it, and subsistent livestock and wildlife. Human exposure to nuclear and chemical tests and factories, or via the food chain, results in miscarriages, malformed foetuses, high infant mortality and congenital disorders, leukaemia and other cancers, tumours, thyroid disorders, and complex debilitating and life-shortening syndromes.[6]

The people responsible for this are our heroes? These people are why we take holidays and sing anthems? What a maddening concept.

The strangest part of this is that media whores call people in radical environmental groups "eco-terrorists" for burning down a slaughterhouse or sabotaging deforestation equipment, but not if those people are dressed in camouflage and just following orders. Think about the military engaging in similar or worse activities. For example:

- The military enabled Hitler to execute anyone he didn't like—not just Jews. What eco-terrorist has done this?
- Want an example closer to home? Iraqi sanctions in the 90s (which are violent acts of war—it's withholding needed food, medicine, etc.) killed a half a million Iraqi children. Our "leaders", who control the military, think that's just fine, since it allows for global capitalism to continue. See for yourself. Type "Madeleine Albright worth it" into a search engine and several videos pop up. Dead kids are a reasonable

price for oil. I can't think of a more glaring example that government officials are psychopaths.

- Speaking of search engine videos, if you want some images that will never leave you, search out military personnel throwing a puppy off of a cliff, shooting a missile into a herd of animals, or a group of thugs beating a sheep to death and laughing about it. Or, you can easily find soldiers randomly firing off land base-destroying bullets, missiles and grenades out of boredom. This is proof of the very obvious psychological damage being in the military causes otherwise normal people.

- Are you religious? Roman soldiers killed Jesus. Once again, it was people just following orders.

- Militaries make all genocides possible. When people talk about governments causing death (democide) they're talking about militaries following orders. In the 20th century, aside from the Holocaust, soldiers directly with violence or indirectly through starvation killed tens of millions of citizens in, just as two well-known examples, Russia and China. Mass graves all around the world are kept full thanks to military power, and, to a lesser extent, the police who "Protect And Serve" whoever is paying them regardless of morality.

Further, the armed forces allow for planetary resource plundering. The military has also allowed for dictators to destroy hundreds of millions of people usually for being in the wrong place at the wrong time. The military (and, locally, the police who are the same thing with fewer guns) stand guard while giant corporations systematically eradicate everything we need to live—clean air, clean water, and unspoiled land. In this savage culture, if someone kicks down the door to a vivisection lab to save animals from torture, the police (and, if needed, the military) will send wave after wave of constumed thugs to neutralize that unarmed eco-terrorist by any means necessary.

If your idea of "freedom" is a planetary wasteland destroyed by industrial capitalism and war—which are the same thing—continue praising the soldiers who are doing it. Just hire some people to change the language. Call them "heroes" and label anything they fight "terrorists."

How is this heroic? How is this even sane? Apparently it is actually insane considering a military person commits suicide in the United States at the rate of about one per hour. It may actually be more than that since the only statistics collected are from those service people who are in the Veterans Affairs system.[7]

Again, I don't want to come across as somebody who is condemning everyone in the military. After personally having numerous conversations with veterans it is very apparent that their government and especially their recruiters lied to them. I'm sure many think they're doing the right thing and serving their country. I don't want them or anyone else to be in the military industrial complex. I just wish all those who are serving would realize they could stop this culture's suicide march simply by not obeying. If they don't, after they destroy the planet—then what? How would they even explain this to their children? What would that conversation sound like?

"Dad? What happened here? I researched the military since you were a soldier. It says the enemy terrorists attacked us and we fought back, but the fighting never seemed to stop. I also looked into bullets, bombs and missiles the military used on the terrorists, and I found out they have a lot of deadly chemicals that poisoned our land and rivers. Is that why we have so little food and water? Because of the weapons?"

"Yes, son. War is hell. But we did what we had to do."

"But dad, doesn't that mean it's kind of your fault that so many are sick and dying right now?"

"Son, it's not my fault. I was just following orders."

Notes:

1) Eight Out Of 10 Americans Support War In Afghanistan:
http://www.gallup.com/poll/5029/eight-americans-support-ground-war-afghanistan.aspx
2) American's Staggering Defense Budget In Charts:
http://www.washingtonpost.com/blogs/wonkblog/wp/2013/01/07/everything-chuck-hagel-needs-to-know-about-the-defense-budget-in-charts/
3) Number Of US Military Bases Worldwide:
http://www.lewrockwell.com/2012/02/laurence-m-vance/fact-checking-wapo/
4) One Nation, Under Water:
http://youtu.be/JIi_a93uaa8
5) Effect Of Exxon Valdez Oil Spill on the Economy:
http://useconomy.about.com/od/suppl1/p/Exxon_Valdez_Oil_Spill_Economic_Impact.htm
6) War And The Environment:
http://www.ppu.org.uk/learn/infodocs/st_environment.html
7) Why suicide rate among veterans may be more than 22 a day:
http://www.cnn.com/2013/09/21/us/22-veteran-suicides-a-day/

12. Love

If you're reading this book cover to cover, or if you've read other chapters, you might think that I am an angry, hateful person, or that I hate people. I only assume this because I hear it just about every day, along with words such as "pessimist," "misanthrope," "nihilist," and my personal favorite, "curmudgeon."

None of this is true. Even if every name people called me were accurate, it would not change the fact that civilization is destroying the planet and threatening everyone on it.

Learning about the brutality and extent of civilization's hatred for itself and everything else is heartbreaking work. I've been studying for years but recently I think the information has truly sunk in, which is why I wrote this book. I see horror everywhere, all the time—even in my dreams. During a typical day, it is impossible to shut it out.

I sit at a desk; some people call it "wood." I call it what it is: dead tree bodies. Logging. Clearcuts. Torture. Destruction. Animals displaced or killed. Rivers choked out from sediment. A graveyard where a forest once stood.

I'm typing all of this out on a computer. Computer? You mean stolen oil, third-world exploitation rare Earth minerals, and burned fossil fuels spewed into the air at every step to create and ship this machine?

Can I hide from it in my house? House? You mean 8,000 lb of waste[1] on stolen land that people took by force, destroying anything in their way?

Sometimes I want to go out to eat at a restaurant to not think about this stuff for a little while. Restaurant? You mean a location where animals and trees were destroyed for capitalists, now serving factory farmed dead bodies from equipment made by dominating nature? Little kids in poor countries went into mine shafts to get the metals to make the oven that made the food on the fork now in my mouth.

It just goes on and on. I wake up, pick up my phone made in a slave labor country, check the time because this entire culture runs on schedules, put on my third-world labor sweatshop clothing, eat my factory farmed food, get in my polluting automobile and go to work to make money for bankers and the government—the two worst offenders in our current planetary destruction via business and the military? Maybe I'll leave work later and come back to my 8,000 lb of waste and sit in front of the television, since denial and escapism are easier to deal with for a couple of hours. I'll fall into a fitful sleep and then get up and do it all again. By participating in this system, my hands are blood red. I try to change things, but I can never do enough for the oppressed. Being alive in this culture means I am complicit in violence, waste and daily, invisible torture and massacres.

It all really screws with my mind. I don't see "things" anymore; I see the compacted end results of cruelty. I am finding it increasingly difficult to enjoy anything since while I am enjoying something my fellow Earthlings are suffering and dying. Even reading books like this makes me think about the trees taken down to make them and the pollutants now in the atmosphere and water to ship them.

Daily, I debate, write, share articles and talk to people. This helps a little but people who don't know or care about their world are having children every second of every day. They are then passing on the values to their children that it is okay to do whatever they want to nature.

Every day, I see self-important people all around me who feel really good about the fact that they are "public servants" or capitalists or whatever. They seem to think that they are doing great work and improving people's lives. They feel entitled to supp at what's left of nature via whatever justification they can come up with: God, where they were born, their species or their social status. The conditioning is strong. It seems to get worse by the hour. I can see it unfolding. I have a front row seat. No matter how much I want to look away, I can't. I won't. I turn and run but never make it very far.

I had to go to a furniture store last night for work. I actually had to fight an anxiety attack to stop from screaming out loud.

"THIS PLACE IS FULL OF BODIES! DO YOU NOT UNDERSTAND THIS!?"

It was row after row after row of dead trees, forcibly extracted metals, barely conscious humans, children who will be raised to hate nature, plastics torn from the Earth...

I got what I came for, left as quickly as possible and went home. I went to bed but could not sleep. I stared at the ceiling for a long time wondering what the hell I could do about this. It affected me almost as much as the first time I saw slaughterhouse footage. Big box stores are the same thing--houses of slaughter where you pay for the service of someone else committing violence for you. (The secret of success is to keep torture invisible and keep the bottom line in the black.) No matter where you go, whether it is a furniture store, the supermarket, a car dealership, or wherever, everything you see in every direction got there because of violence. There are almost no exceptions. Even locally grown and produced goods can be counted since we are all living on stolen land based on a global economic system.

What can I do to stop the furniture stores from rolling over old growth forests? I will be one man outside alone with a sign. There's no group environmental protesting formally because the trees can't scream. (I actually think they can, but hardly anyone listens.) Those organizations are too busy making compromises: "Okay; you can kill half the forest but leave the rest alone!" Ten years later: "Okay; you can kill half the forest but leave the rest alone!" Average people will give money to these organizations and feel like they are part of the solution. What sense does that make? Would people have given to the Nazis if they had only killed half the Jews?

Is this all for real? Is this my life? How can this really be happening?

I have a secret fantasy that all of the UFO rumors people often speak of are true. Maybe you've heard of one in your area. If you see one, please let me know. They might be here to pick me up.

Sadly, this won't happen, so I will never stop speaking out. Why? Because I love the Earth. I love the trees. I love the rivers. I love the people. Through my angry writing it may be hard to tell that I love the people, but the truth of what is happening to them and what they are participating in is very ugly. I cannot gently explain that their actions and inactions are helping kill the Earth and that their children will have to deal with the consequences once their parents are dead. It is hard for me to keep calm when every day more and more species go extinct, more and more old growth forests are destroyed, more and more rivers and oceans are polluted, more and more effluence gets into the air (and subsequently into our bodies), and more and more Earthlings suffer and die for a dead-end, insane unsustainable culture.

I have become accustomed to mourning every day. I feel the weight of my actions and inactions constantly. I think always of ways to stop it, of how to protect my immediate family and my extended Earth family from it, and when I go to sleep, I dream about it. I consciously tell my mind to think about it while my body rests. I need to find the answers. I need to make it stop.

Friends and partners have told me that it seems like I am "mildly crazy" because I dwell on this information. That may be true, but I feel like if I dig deep enough, if I learn enough, if I act enough, if I tell enough people, maybe somehow it will stop. I feel like this is very sane—protecting what I love. Insane would be to let it continue without caring. I will always care until I take my last breath. If there is an afterlife, I will continue the struggle in whatever form I take. For now, I am fully awake and fully alive. Although the work is hard, it energizes me like nothing else. I can never stop.

The only conclusion I have come to about why I can't go back to sleep and let the world burn is because of love. I love it all. I want it all to be safe and free. I don't want evil people wrecking it because I love those humans and non-humans who haven't been born yet. I want them to have a fertile, beautiful planet to live on. They deserve it just like we do. And since we were born out of the Earth, it is like our mother. We can't live without it. And this culture is systematically butchering it for short term profit.

Isn't protecting your Earth Mother sane? Isn't guarding her children worth your love? They are all running out of time. If you love like I do, there do something before it's too late. There is no more noble cause.

This culture hates you. It hates me. It hates itself. For shit's sake, it frowns upon us and makes us ashamed of our fucking bodies. Religion teaches men to be mortified of their penises. Civilization invented bras to teach women that their nipples are not to be seen, and that their breasts are for sex, not nursing children. It teaches that little boys must act one way and little girls must act another or the world will shame them. It teaches us that trees exist to be cut down and that life is expendable for petty, human-centered reasons. It teaches us that we are separate from nature and that we can and should dominate it even to the point of destroying ourselves. It alienates us from each other and the world. Why would anyone have allegiance to it?

I guess we all have to find our own reasons to fight. Maybe I haven't convinced you with my experiences, but I want to leave you with this: do you have a cat or a dog, or have you known and loved one in your life? If yours is around you, take a second to pet him or her. Look into your friend's eyes and take in just how much he or she means to you. Feel that love? Good.

Now, realize that every 11 seconds in the United States, a pet much like yours is murdered in an animal shelter. 2.7 *million* of them.[2] Imagine taking your purring cat or your happy dog with his or her wagging tail, leading them into a room, putting him or her on a cold, steel table, and then shoving a needle into a paw. Imagine watching your friend collapse and die. Or, imagine the more cost effective method of throwing happy, healthy animals on top of each other in a box and gassing them to death. I have seen this happen. I have heard them scratch and scream. I have seen their lifeless bodies. It made me physically ill.

That is what capitalism and civilization are doing to pets just like yours—often even the same breed. If you have any empathy, after learning this, you may be feeling sadness and anger. That is a sane reaction to insane information. That is your conscience, your love. Take a moment and let yourself feel sad. It is completely normal. Now think—this happens all around the Earth, every second of every day. This is what civilization does. Why doesn't anyone do something about it or even seem to care?

Don't you want the pain, misery and execution of innocent animals like your friend, your four-legged family member who you love, to stop? Don't you think that is worth fighting for?

Notes:

1) Construction Waste Recycle:
   http://constructionwaste.sustainablesources.com/
2) Pet Overpopulation:
   http://www.humanesociety.org/issues/pet_overp
   opulation/

13. Manifest Destiny

If you want agriculture, you first need to find an arable piece of land. If there are creatures on it, they will need to be removed. Quickly now; it is your right. A book written thousands of years ago told you so.

To effectively work (exploit) the Earth, you need metals.

If you want metals, you have to have a mine. If you have a mine, you have to hire security to keep poor people away from it and to keep workers from stealing from it. (Look around; the police stand in front of everything somebody needs.) Mining is miserable work, but don't worry; employees will be abundant since, after destroying their way of life via agriculture and domestication, they will have no choice but to work in the mine. Bosses will keep them in line. Hierarchies develop. This is the way it should be. Plus, in a few generations, the workers will forget that they were once independent and call their current situation, "just the way it is".

Agriculture means more food at first, so more people will have children. That means more agriculture, which means more metals. More smelters. More security. More houses built. Take more land to build those houses. Kill off nature or animals on the surrounding lands. Cut down the trees. Clearing the forests requires more metal. More mining. More workers. More expansion. More force. (If the people on the land with the resources you need continue to resist, you'll have to "convince" them.)

More resources now available means more babies, which means people spread out, which requires more transportation, which needs more infrastructure. That requires more metal. More workers to do miserable work. More security to guard the infrastructure, agriculture, smelters, property and mines. More exploitation of nature. More death. (Just don't kill each other and there is no problem.)

Eventually you will run into more people and animals who don't want to give up their land. They will fight back. You now need an army. That means taxes. Tell the people it is for their own good. It will make them more secure. It isn't offense; it's defense! You will need politicians to convince them. Politicians are your friends. They look out for you. Without them, your enemies would destroy you and your family. You want to protect your family, right?

You need to create fictional justifications of why people are better off domesticated under the new agriculture system. Call it "progress" and demonize the ancestors who lived under the old way. (You don't want anyone not dependent. That would threaten your new, better, deserved way of life.) You'll need the leaders to repeat this over and over again until people believe it. Tell all the children. Get them young. They are the future. This way is the best way, the only way.

Tell the workers that they only have to be nice to each other. Don't aggress. Follow the rules and you'll get ahead. Nature is not special like us. It is here for us to use. The children are starting to agree. They will tell their children.

While taking what is yours, you will encounter rough terrain. You need technology--chainsaws, ploughs, etc. More comes out of the Earth. The humans living on the land with the needed resources don't want to give it up. Send in the military. Create propaganda to justify it. More leaders emerge to tell the new generations that this is the way it should be. They hire more enforcers to keep people in line. Without that security, the system breaks down.

This is normal. It is for your own good.

More layers of control develop. People spread farther out. We need regional leaders. We need more land. More resources. That means more workers. The workers are now hopelessly dependent on the system that exploits them. Any that speak out need to be dealt with. That requires police, plus leaders to tell everyone that it is all okay. "Nothing to see here. Move along. Keep your head down and keep quiet. You can't make a difference anyway."

The military makes capitalism possible. Without their guns, we wouldn't be able to get or transport the raw materials to make stuff that makes our way of life possible. Capitalism is the best way--the only way. We wouldn't want to go back to how those "savages" lived, would we? The ones who created technology to preserve their land instead of destroying it? That would be ridiculous. After all, then we wouldn't have smart phones, indoor plumbing or fast food available twenty-four hours per day. And *everybody* knows that is the only way to live. The leaders told me so—my friends, the politicians.

It needs to expand. It must keep going. More. *More*. Never mind your soul. Never mind your planet. Never mind your children's future.

We are watching you now. We know what you are doing at all times. Do not get out of line. It is for your own good. Those eco-terrorist environmentalists are just trying to scare you. We have your best interests in mind. Listen to us. That's it.

That's right.

Go back to sleep now...

14. Nationalism

I'm going to come right out and say it: nationalism is absurd. It has caused little more than false separation and gratuitous conflict among poor and minority humans. I can sum it up with a simple sentiment: "we're better than you," and you can imagine what comes next.

Nationalism's actual definition is:

loyalty and devotion to a nation; especially : a sense of national consciousness exalting one nation above all others and placing primary emphasis on promotion of its culture and **interests** *as opposed to those of other nations or supranational groups*[1]

The emphasis on "interests" is mine; we'll get to that shortly.

Like property rights and ownership, nationalism is human-created fiction. It is used by one group to control another—going as far as murdering native peoples and destroying land bases for the glory of a country. (Keep in mind that countries and borders are also imaginary.)

An example: "In 1492, [Christopher] Columbus sailed the ocean blue." Yes; I suppose that's true, but why he did this is not part of the rhyme. He sailed the ocean blue as a conquest to pilfer resources from foreign lands. Why? Because the monarchs of Spain financed it. Spain was at war with Muslims and was already stealing land, so why not send out Columbus to possibly steal more—especially, so they thought, in China and India, which were both rich in resources.[2]

Columbus landed in the Bahamas on his first voyage and claimed the land for Spain. He just made that up. He literally got there and made something up, which was enough justification for his men to follow orders. He had no authority to claim ownership of anything especially considering the Arawak Indians and many non-humans already lived there.3 Of course, he and his crew mercilessly enslaved and massacred the local populace, but that's okay because it was for Spain (whatever that means).

This is one of history's most famous examples of nationalism and it was repeated over and over in the Americas. After Columbus set precedent, the Europeans followed and spread like a plague, claiming ownership of every inch of land in North and South America—carving it up among the wealthy elites. They did not do this nicely; they slashed, burned, raped and murdered their way from coast to coast. I could cite sources for this but it is common knowledge. I encourage everyone interested in this topic to read *A People's History of the United States: 1492 to Present* by Howard Zinn. It is quite eye-opening and disturbing and, unfortunately, accurate. You will be horrified at the public education's historical whitewashing that your teachers spoon fed you. It really is all lies.

In contrast to what I learned to as a child, it is now my understanding that indigenous people do not view ownership as modern Western civilization does. When they say it is "our" land, it is like if you were to say it is "my" mother. You don't own your mother, but she is your kin, and you show allegiance to her. Like the land, nobody should be able to lay claim to, buy or sell your mother. (Try telling that to a capitalist.)

I wish I could say that nationalism and all of its horrific consequences no longer exist in the twenty-first century, but I can't. I'd be lying. The "Western" lifestyle, which has a never-ending urge for more, continues to chew up the Earth and displace indigenous peoples.

Let's look at a more recent example: the Amazon rainforests. Survival International reports that "[t]he Amazon is the world's largest rainforest. It is also the ancestral home of 1 million Indians. [I can't believe we still call them that.] They are divided into about 400 tribes, each with its own language, culture and territory."4 Of

89

course, there are countless non-human plants and animals living there. The proper thing to do is to leave them be, isn't it? After all, it is their land. They need it to live. Right? Wrong, says anyone culturally indoctrinated into civilization.

Civilization created something called "government" that claims that certain humans own everything. Nationalism is of course a part of that—meaning people in a specific area need to claim allegiance to the government's interests. In other words, the government decides who owns what, and the citizens who go along with those edicts are called "patriots."

This all sounds okay, but the problem is that word: interests. ("Interests" is a euphemism for "wants.") In almost every government, the wealthy elites are the ones who make decisions for the citizens. Their interests are rarely in line with the citizenry—especially the indigenous and non-humans. Nevertheless, the military enforces the government's interests and puts down resistance by any means necessary, despite the fact that the vast majority of soldiers are part of the exploited class! (I provided detail of this in a previous chapter—see "Killers".)

The elites will use the media to promote the idea that nationalism equates to the government's agenda. In other words, you're a bad citizen if you don't agree. Your country *is* your government's interests. You are unpatriotic if you dare speak up of (gasp!) resist. The entire thing is nonsense; it's just something somebody created out of thin air. It blows my mind that anyone believes it.

Nonetheless, people can be convinced that beliefs like nationalism are good and deserve action. Unfortunately, putting one culture above others—in other words, believing in nationalism—creates hierarchies. If the American way of life is better than others, then it's okay for the US military to invade other people's countries. It's okay to topple foreign governments—including and especially indigenous people and non-humans. Nationalism makes it okay to clear cut foreign forests, to dump toxins in their rivers, to use the military to force third-world citizens into poverty, to take natural resources, and to commit violence and atrocities around the globe. In short, nationalism is a carte blanche excuse to enslave and exploit anyone and anything not within one's borders. Just make sure you have the media stand up and applaud every act of violence, plus couch it all in euphemisms like "spreading democracy." They hate us for our freedom, right? It's definitely not because we blew them up and took everything they had.

Of course, a natural extension of nationalism is to brutalize citizens in one's own country who do not agree with the government's interests. Just convince the general public that laws are legitimate and label anyone a "criminal" or a "terrorist" who disagrees. (The media is invaluable here.) If the government decides to murder old growth trees, set up factory farming in your backyard, to deprive you of food, or whatever sick interests the ruling class is pursuing, your resistance could lead to your death.

Regrettably, there seem to be plenty of people willing to kidnap and/or kill you if you step out of line. I just did a search engine query for "activists arrested" and it came up with over sixty million results. Activists killed? 117,000,000 results. One of the top hits for this was an article titled, *Environmental activists 'being killed at rate of one a week'* from the London Guardian.5 The content is not new or surprising; it's war over natural resources in the Amazon rainforest.

Civilization is willing to endlessly murder people to keep itself going, and there's no shortage of nationalistic justification. It is patriotic to kill resistors. If that was not true, the government would promote and assist activists rather than caging and butchering them. Since they don't, the only logical conclusion I can think of is that nationalism and patriotism are okay with widespread violence so long as the home country gets what it wants. This ludicrousness is nothing more than fiction to support the interests of whatever sociopath or psychopath is in charge.

Don't believe it. No matter how many people tell you the lie of nationalism, don't believe it. People believing and repeating the lie doesn't make it true. It makes those people insane. (It may also mean that person has nothing to be proud of, since being born in a place is a genetic accident, not an accomplishment.)

Your allegiance is with the planet that gave you and sustains your life. Without it, you would never have existed, and if patriots destroy it, you will die with it.

Notes:

1) Nationalism Definition:
   http://www.merriam-
   webster.com/dictionary/nationalism
2) Christopher Columbus Biography:
   http://www.biography.com/people/christopher-
   columbus-9254209
3) Columbus, The Indians, And Human Progress:
   http://www.historyisaweapon.com/defcon1/zinnc
   ol1.html
4) Survival International/Amazon Tribes:
   http://www.survivalinternational.org/about/amaz
   ontribes
5) Environmental activists 'being killed at rate of one
   a week':
   http://www.theguardian.com/environment/2012/
   jun/19/environment-activist-deaths

15. Order

I have in no way felt superior to animals or plants. Ever. I simply never believed it—even in my childhood. I grew up as a Christian, yet I never understood nor paid much attention to Genesis 1:28, which says,

God blessed them and said to them, "Be fruitful and increase in number; fill the earth and subdue it. Rule over the fish in the sea and the birds in the sky and over every living creature that moves on the ground."[1]

Even if you are not of that particular religion, there are other historical justifications telling humans they are superior to all else. For example, extrapolated from works by Plato, Aristotle and Proclus is the Great Chain of Being[2], which puts God at the top, then Man (never woman, of course), Animals, Plants, and then "inanimate" objects like rocks, Earth, etc.[3] (This of course disregards the fact that soil is rich in living creatures like bacteria.)

Even if I were to completely disregard any and all religious and historical writings concerning supposed natural hierarchies, I would only need to look at the way humans have treated the Earth for the last 10,000-12,000 years—since they began to domesticate plants and animals. Humans have warred over the Earth, all claiming it belongs to them for this reason or that. They have ruthlessly torn down trees, dammed rivers, taken great rocks and precious metals to forge into weapons and machines, and have exterminated any indigenous people, plant or non-humans in their path. Great cities and civilizations have come and gone under the belief that somebody, somewhere, owns all of this and has the right to take what they please.

With that in mind, one can easily see why civilization happened and why it continues to exist despite voices throughout history calling for change. Cultural straw man arguments seem to make sense. For example, if you came and took my wallet, I would run after you to try to get it back. Therefore, on a larger scale, if you took "my" land, I would fight you to regain control. I might even cut down all of my forests and dam my rivers to acquire the means necessary to take back what is mine, plus, since I was able to beat you, what you claim is yours. Civilization and capitalism say "might is right."

The problem with all of this of course is that nothing belongs to us. We came from the Earth, and when we die, the soil will consume our bodies and transform their building blocks into something new. Given enough time, none of us can escape that. From the tallest mountain to the most vicious human, we will all return to the Earth eventually.

The notion that humans feel they are "above" everything else is absurd. It would be like if one of my ears rebelled against the rest of my body. The ear would then recruit the nose, and then maybe some fingers. After a while, the toes might rebel against the fingers. They would ravage the legs—each claiming ownership. All the while, the nose and ear would poke out the eyes and fill in the mouth in order to defeat each other.

In the end, it would all be one giant dead zone where no one could survive. This is the direction civilization is headed in very, very rapidly. Since humans' lives depend on the Earth, Air and Water they are all destroying in order to assert dominance over one another, after all is said and done, no one will be alive to rule the land.

Did you ever learn anything like that in school? Probably not. I learned the status quo order of things: that government rules people, people called "farmers" rule the land, trees are kind of in the way and should be cut down to make room for hair salons, and that sure, slaughterhouses did exist, but it's okay because they are just killing "animals." Then, my teachers called humans like Hitler and Stalin "animals" for committing their numerous atrocities. Even in popular culture, every criminal is an "animal". For you see, animals are simply violent brutes, and we are above them.

Since animals are violent brutes and we are above them, my teachers said, it is perfectly acceptable to rule them. If only animals could all walk upright, dress in clothing and drive cars, then they would be in such a position that ruling them would be immoral. For now, just look at them! Those animals are all out there killing each other for food and wallowing in filth.

Yet, a lion eating a gazelle is not a hierarchy, nor is it a situation where one is greater than another. A conditioned person might try to make this argument. I, on the other hand, want to highlight it as untrue. Humans project hierarchies and order into nature where there is none. Lions do not capture gazelles and put them into factories for wage labor. They kill and eat them to survive. There is no philosophy behind it. Anyone who tells you there is has been culturally conditioned. Gazelles and lions are just different, not better, worse, or deserving of rule.

Further, find any trait that humans have and I would be willing to bet some animal somewhere has it or can do it better. Some of the more well-known "superior" characteristics are bulls being physically stronger, cheetahs running faster, birds flying without airplanes, and gorillas capable of learning sign language *and* speaking in their native tongues. (Please find me a scientist who can speak as much gorilla as a monkey can sign.) Koko the gorilla will even teach you sign language.[4] Imagine that. Brutish beasts teaching us things!

To me, it seems common sense that if some humans put themselves above animals and plants, they would eventually find justification for putting themselves above other, vulnerable humans as well. Maybe humans who speak a different language. Maybe humans who are from another part of the world. Maybe indigenous people living on land prior to someone else. Remember, you cannot conquer and enslave them if they are your equals. No; they are "animals" and we would be doing them a favor teaching them to be like us.

Here are a few prominent examples where humans used this natural order justification to get what they wanted:

a) Slavery in the United States: Proponents—usually rich plantation owners—cited many reasons for owning other human beings. They took evidence out of the Bible, looked at historical cases of slave ownership, and even used the legal system to assert that black Africans were not people, but property.[5]

b) The Holocaust: I do not want to create a simplistic argument here, as the Holocaust happened for many complex reasons. One of them was, however, a cleansing for anyone outside of the master Aryan race. The theory was that Germany had become weak by interbreeding with inferior non-Aryans. [6] Aryans were therefore better, or above others.

c) United State Indigenous People: Settlers commonly referred to these vulnerable humans as "savages" who needed enlightenment. Even intellectuals like Thomas Jefferson agreed. "[He] believed that if American Indians were made to adopt European-style agriculture and live in European-style towns and villages, then they would quickly 'progress' from 'savagery' to 'civilization' and eventually be equal, in his mind, to white men."[7]

Civilization simply would not be possible without hierarchies and people willing and able to violently enforce them. It takes a lot of time and effort to maintain superiority and there are examples everywhere. Drive by an abandoned house and see how different it looks then one that is maintained. Liberate a piece of land and watch nature take it over and rewild it. Pets, who are our prisoners, will become feral if let outdoors. The tallest building will crumble back to Earth if left alone for, geologically, the blink of an eye. Even humans will stop conforming if other humans stop putting guns to their backs.

There is no order. The entire concept is fictional and is being used to control you and your family. If you realize this, you can begin to re-educate yourself to the fact that the Earthlings are all in this together, and that we all must depend on each other for survival rather than bowing to a sadistic fiction that claims one species rules above all. The concept of natural order is nothing more than propaganda that people who want to rule others created in order to control them. Stop believing it and you can start to be free.

Notes:

1) Genesis 1:28:
   http://biblehub.com/genesis/1-28.htm
2) Great Chain Of Being:
   http://academic.brooklyn.cuny.edu/english/mela
   ni/cs6/ren.html
3) Great Chain Of Being Origin:
   http://www.blackwellreference.com/public/uid=3
   /tocnode?id=g9781405106795_chunk_g97814051
   067958_ss1-58
4) Learn to Sign with Koko:
   http://www.koko.org/world/signlanguage.html
5) The Southern Argument For Slavery:
   http://www.koko.org/world/signlanguage.html
6) The German Master Race:
   http://www.schoolhistory.co.uk/gcselinks/source
   work/masterrace.pdf
7) American Indians:
   http://www.monticello.org/site/jefferson/america
   n-indians

16. Property Rights

In my many conversations with anarcho-capitalists, they have told me that I do not respect property rights. I just don't get it according to them. (Also, I'm a communist and a kook, but I guess that goes without saying.)

Let us look at this subject in depth. According to *The Concise Encyclopedia Of Economics*,

A property right is the exclusive authority to determine how a resource is used, whether that resource is owned by government or by individuals.[1]

Let's break this down piece by piece.

First, property is a word humans invented to justify ownership. Both property and ownership are fictional concepts. The only way a person can prove ownership is with a deed or another piece of paper, which is, again, fiction—only this time it's legal fiction. Alternatively, a person can say they own something or someone, but—you guessed it—it's still fiction. Everyone has to agree that somebody owns something for the piece of paper or verbal claim to be valid. Nevertheless, it is fictional; just because a lot of people believe something does not make it true or correct.

If you think about it, we "own" nothing. Not even ourselves. The Earth owns us. We came from it and to it we will return. Everything we are and everything we need to survive comes from the land. The land does not need us; we need it, so we certainly cannot own it—especially if we just showed up a lifetime or so ago. Creatures have been on "our" land long before we existed and, unless we inflict capitalism on them, they will be there long after we're gone. The land has been and will be there even longer.

Some economists like to equate ownership and property rights with human rights, but, now continuing to the second word, "rights" are fictional. The generally accepted definition of a "right" is that which is morally or socially correct.[2] We're treading now into the realm of philosophy, which is a school of thought humans created that projects their perceptions into nature. It is fine mental gaming, but like "property" and "rights," "philosophy" is fictional. To prove this, please try to find a polar bear or goldfish that acknowledges and/or agrees with Socrates.

"God-given", "natural", "Constitutional" and all other adjectives preceding the word "rights" mean nothing; it's all human fiction. At its core, a "right" is a system of control. If rights can be given, they can be taken away—so long as you can convince people that deprivation is moral. It's easy to deprive others of rights or exploit them if you just make some stuff up. Conversely, if any right was inherent or self-evident, nobody would need to talk about it. It would be like creating an entire system of often contradictory beliefs to explain why the sky is blue.

Exclusive authority? Now we're getting to the meat of property rights. Authority is, again, something humans created to justify why they can do what they do to the natural world. If I give somebody money (fiction) for a deed (fiction) for a piece of land (the designation or naming of which is fiction), I then own (fiction) it and can, according to property rights (fiction) do as I please with it. Isn't that convenient? I suppose if you're a wealthy human this works out really well. Too bad for anyone else that already lives there.

As an aside, "authority" is contradictory to anarchy. Anarchy by definition is anti-authority. How then can an "anarcho-capitalist" even exist? In that system there's so much emphasis on who owns and controls what. I guess it's necessary to overlook a little bullshit when the system itself is based on the fictional concept that the entire Earth can be given dollar (fiction) amounts.

Now, "to determine" is interesting. What if I really like basketball and hate the woods? I buy up a piece of land and pave over it. What about the non-humans that live there? Why don't they have property rights?

To further complicate land ownership, when I purchase property, do I own the land all the way down to the center of the Earth? What about in the other direction? How far up do I own? All the way to space? Can I do what I want underground or in the sky? Do I also own the birds that fly and the creatures that crawl underneath my land?

What if I want to sell the land to an oil company so they can practice hydraulic fracturing? Do I need to take into consideration the surrounding areas? For example, what if that fracking damages a water table underground that destroys an indigenous community nearby? Since the toxic sludge of bitumen oil is now on those people's land "over there", is that my problem? I can do what I want on "my" land, can't I? What if I allow pesticides to be sprayed in the air on my land and they end up giving somebody's children cancer downwind? Is that my problem?

The whole thing is very confusing. I guess that's why somebody invented the modern legal system (even more fiction).

How a resource is used: I hate that. I *really* hate that. Resource, I mean. It breaks down whatever we're talking about into monetary worth—a fictional human concept used to determine how something will be treated or consumed. If I buy a piece of land with many trees, I may see those trees as resources. Therefore, since I have a piece of paper that says I have the authority to determine how my resources are used, I may choose to murder those trees and clear that land for some banal purpose. It's my... *right*. I can so I will.

I've broken down the entire capitalist system into the following simple idea: "might is right!" Suffice to say that if somebody sees trees and non-human animals as resources, they're bound to see vulnerable humans as resources. Maybe they'll even own them. Maybe they'll create a religious book that justifies owning them. (Why not if we're making things up, right?)

The last part of the property rights definition mentions ownership, government, and individuals—three concepts that are completely fictional. Ownership we've already explained, but government? That's just one group of people who makes up laws (fiction) to explain why they have the right (fiction) to rule other humans and non-humans. I've heard the term "ruling class" thrown around for government, and I think that is appropriate. Those people have more fictional money and fictional power than everyone else, so they have the right to rule. Why not—especially if they use those same justifications to rule the natural world? Thus, hierarchies were born.

Individual? There is no such thing. Our bodies are an amazing system of symbiotic relationships, but we can live without some of them. If I lose an arm, a kidney, a piece of my liver, a leg, and an eye, am I a lesser individual than someone who has those parts? Conversely, should I value my heart and brain much more than my big toe because I need them to survive? Is my heart then an individual? Is it superior to my lung because it keeps me alive?

Further, if a man has a much larger bone structure than I, is he more of an individual? Also, if we truly were individuals, why would the government make laws on how we treat our own bodies (drugs, banned food additives, etc.)?

I can dig this hole much deeper but hopefully you understand that these absurd notions of property, ownership and rights are imaginary—created by people obsessed with control. These concepts were unknown to our pre-civilization ancestors because those people saw themselves as part of a whole rather than as controllers of land and their fellow Earthlings. That is a much healthier and respectful way to live, and it does not deprive other creatures from existing.

I don't think it's any coincidence that humans are the only species that have to pay to live on a piece of land. What other species would want to make up something so outrageous and spill oceans of ink writing about how wonderful it all is?

Capitalism and property rights deprive most creatures of everything including their very lives. Please—count me out. Call me whatever –ism you want, but that system has shed too much blood for me to ever consider it viable, let alone worth any struggle to support.

The Earth is yours, mine, and everyone else's, and we belong to it. Living on it is our "right" because we were born on it, not because we have economic value or property. The sooner we spread this information, the sooner people will realize that denying any creature of its birth right is cruel and unnatural. Carving it up into slices and writing books about why that is good only benefits the rulers. If you care at all for your fellow Earthlings and the planet itself, please join me in openly mocking this disgusting system of control before it kills us all.

After all, if you can deprive someone else with your fictional property rights, believe me that someone will eventually deprive *you* of yours.

Notes:

1) Property Rights:
   http://www.econlib.org/library/Enc/PropertyRig
   hts.html
2) Right Definition:
   http://www.merriam-
   webster.com/dictionary/right

17. Quiet

When is the last time you heard silence? I mean, complete silence? I've tried hard for many years to find it and have been largely unsuccessful. Give it a conscious try some time. It's surprisingly difficult.

I've lived in a bunch of places in the United States, visited many more, and been to still more places. Everywhere I go, I hear civilization's incessant reminder that it is there and that it is in control.

Anywhere I travel where there are roads, I hear cars. If I go into the woods to try and think, I hear chainsaws and off road vehicles. Down at the beach, I hear radios and barbecue grills. In empty fields, I can hear cars in the distance and nearby factories turning living things dead. Even under water, I almost always hear somebody's damn power boat.

Quiet is another casualty on a long, long list that civilization has robbed from humans and animals. Most people know no other way and probably would never stop to consider a less noisy world. They may even come to think that they are entitled to their very noisy way of life even if it affects other animals—and it does. Let me show you what I mean.

I came across a court case called, "National Security vs. Whales.[1]" I think after I read that I had one of those cliché movie moments where I dropped what I was holding. I simply cannot believe that those four words exist in that order outside of some absurd joke.

According to the Huffington Post, "The Navy wants to install an undersea array of cables and sensors for training warships, submarines and aircraft about 50 miles off the Atlantic coast of southern Georgia and northern Florida. Environmentalists have sued to block the project, saying it's too close to waters where right whales migrate near shore each winter to birth their calves.[2]" To most, that probably makes a lot of sense and seems compassionate, but I have a few problems with it:

a) "The Navy wants..." Wrong; the ruling class wants. The Navy is fictional. It's just a bunch of people in costumes following orders who apparently don't know any better. Unless the next words are, "...to stop existing" or "...to dismantle civilization" you know it's going to be bad.

b) "..array of cables and sensors..." So, they're not only going to invade marine life habitats, but they're going to rape and pillage nature first to create these things. Some would call this insult to injury.

c) "Environmentalists have sued to block the project, saying it's too close to waters..." That is very troubling. Environmentalists, by doing this, are implying that project, under certain circumstances, is permissible—just not in that particular water. I can't get behind that. Maybe something like, "...saying it's outrageous that a bunch of warmongering psychopaths want to build even more weapons and military infrastructure." *That* would be environmentalism. Since they do not criticize the Navy and the testing range outright, these environmentalists are complicit to whatever damage the Navy will do elsewhere.

d) "...where right whales migrate near shore each winter to birth their calves." The right whales are indeed right, but this sentence is more whitewashed propaganda. Try, "...where right whales live and have lived for millennia. What 'right' do humans have to invade their space for what amounts to rich white men continuing to destroy the planet?"

The noise from this military abomination, say the environmentalists (I'm going to rewrite for them), would interfere with the right whales' sonar communication. Since there are fewer than 500 whales left in existence thanks to other military abominations and things military abominations enable (like civilization), this is of great concern to environmentalists. It concerns them so much that they wrote it down on a piece of paper and submitted it to the government that owns the military, which I'm sure will convince the military to halt all similar projects immediately and tear down any others that have wrecked habitats.

As I said earlier, I simply cannot believe this entire thing exists. How could any culture be so insane as to think that whales having babies would have anything to do with national security, which of course is another way of saying, "might impede civilization's relentless advance"? All the poor creatures want is quiet and a safe place to live.

Leaving the whales for now, a lack of quiet places also adversely affects humans. According to the Environmental Protection Agency, noise is "unwanted or disturbing sound.[3]" So... everything civilization has to offer? I can't go anywhere without hearing combustion something or other helping this culture destroy the planet. Apparently, I'm not alone; it's actually bad for our health. Says the EPA: "Studies have shown that there are direct links between noise and health. Problems related to noise include stress related illnesses, high blood pressure, speech interference, hearing loss, sleep disruption, and lost productivity."

Along with more quiet, many academics have said that more nature is beneficial to humans—especially kids. Here are a few examples:

- One study showed that children are spending half as much time outside as they did a generation ago.[4]

- This of course has lead to an increase in depression and other mental disorders. One study[5] showed that 1 out of every 8 children is on Ritalin, while another[6] showed an almost ten percent increase of these drugs among kids in just four years.
- Being inside so much is also making kids fat. Obesity from 1980 to 2008 has more than tripled.[7] Of course, the risk of cancer, heart disease and many other ailments balloons along with those waistlines.

Fat, drug-addicted nature-estranged children are going to grow into corpulent, substance-abusing sedentary adults. Then, they'll have more little obese, pill-popping couch sitters and the cycle will continue. Civilization will press on while we kill ourselves and the whales rather than dealing with the obvious problems that are screaming in our ears.

The clear conclusion is that nature is good while being inside all the time is bad. Why? You can find peace and quiet in nature. But not inside. Not anywhere no matter how hard you try. We move from building to building, from screen to screen, from distraction to distraction. (Have you noticed you can't go anywhere without seeing a television or computer?)

People have such impotent "solutions" when attempting to block out civilization's clamor. Have you ever heard of a white noise machine? It's a nifty little thing that plays any number of sounds to help people sleep or meditate. Those sounds are usually something natural, like a waterfall, rain, wind, or a tree rustling—you know, because nature is good for you. Most people, when quieting the mind, do not listen to jackhammers, tractors, car horns, reality shows, or any other industrial sound. They look for quiet, or, in other words, anything other than civilization. (They even plan vacations away from their lives in search of peace and quiet.) And they have lots of options.

There is an entire industry to helping us shut our minds up an off. Pills to help us sleep, classes to help us meditate, the aforementioned white noise machines, isolation chambers, nature "retreats," and so on.

Am I the only one who sees this as bizarre and as something that has a really obvious solution?

Now, if you'll excuse me, I'm going to get the hell out of this chair and room so I can take a walk in the woods.

Notes:

1) National Security vs. Whales:
   http://nsglc.olemiss.edu/sglpj/Vol1No2/Schaffner
   .pdf
2) Right Whales vs. Navy Offshore Training Range:
   http://www.huffingtonpost.com/2012/03/15/end
   angered-right-whales-navy_n_1348606.html
3) Noise Pollution:
   http://www.epa.gov/air/noise.html
4) Institute For Social Research—CHANGING
   TIMES OF AMERICAN YOUTH:
   http://www.ns.umich.edu/Releases/2004/Nov04
   /teen_time_report.pdf
5) Ritalin—Better Living Through Chemistry?
   http://www.dadi.org/rtln_sax.htm
6) *Psychiatric Services* 55(4):387-391—"Trends in
   the use of antidepressant medications in a nation-
   wide sample of commercially insured pediatric
   patients, 1998-2002"
7) Childhood Obesity Facts:
   http://www.cdc.gov/healthyyouth/obesity/facts.ht
   m

18. Resources

I feel like I have beaten this subject into the ground, but it is so prevalent and heinous that I feel I needed to dedicate an entire section to it. Please forgive any repetition, but those who are skipping around this book will appreciate it.

I hate the word "resources" as much as I hate "green", "free-market", and "science". I come across these words in debates and casual discussions almost every time I try to talk to someone about values. In short, if green companies were allowed to utilize science in the free-market, resources could be allocated properly to maximize human potential. (Forget everything else's potential I guess.) This is all ivory tower gibberish. Sit tight and I will explain.

"Resource" is a word humans invented that allows them to feel good about exploitation. It is much like using "animal" in a derogatory manner to lambaste criminals or other undesirables. In other words, if I call something a resource, I objectify it and make it inanimate and therefore it is available for me to do with it as I please. A tree is not actually a tree; it is a resource available for human use. This is industrial capitalism's cornerstone. It posits that everything on the planet is for sale and should be used to enhance whatever pursuits humans choose. Nature is nothing more than a giant department store with low, low prices.

To demonstrate how insane this is, I am going to focus on trees. To most conditioned citizens living in civilization, a tree is a thing that is exists to get cut down and turned into paper or somebody's house. Or, maybe a surfboard. Or a model airplane. Whatever. It does not matter, since it is "just" a tree.

With all the axes and chainsaws in existence, it seems like the human race is doing battle with trees. It hungers to destroy and use them. Capitalism is taking

down 36 football fields' worth of trees per minute.[1] But...
they are just trees, right? It is okay for us to cut them all
down. Wrong.

A tree is a miniature ecosystem. On a tree farming
website, I found a list of those potentially affected by removing trees. Paraphrasing:

Other trees, different plant species, insects, arachnids,
small and large mammals, rodents, birds, and humans.[2]

They forgot, among other things, mushrooms, rivers, soil,
the air the tree might purify, fungi, lichen and other symbiotic creatures.

Cutting down a tree or especially a group of trees
for resource harvesting has a ripple effect. The animals
and plants living in, on and around that tree are displaced
or killed. The soil loosens, which may cause runoff into
rivers, which kills whatever is living there. (Remember,
rivers are alive.) The sunlight bakes the forest floor where
the trees once stood, which damages the bacteria and
makes growing new trees difficult or impossible. And, it
releases huge volumes of carbon into the air that the tree
once held. And for what? Some copy paper? Another
house (even though "[t]here are more than five times as
many vacant homes in the U.S. as there are homeless
people, according to Amnesty International USA."[3]?)
Confetti for a parade, which has a life span of about three
seconds before it ends up in landfills? Obstacles at a miniature golf course?

What hubris! What right do we have to do this?
What possible justification could humans pull out of their
asses to make this okay?

This rampant tree murder is all so trivial and selfish, and it is just one example of the planetary rape
humans in power have decided is normal and culturally
acceptable. Not only that, but the system has to keep
growing or it will collapse, which means infinitely more

113

objectifying, exploitation and murder. If people stop spending money, companies stop making things, and, heaven forbid, industrial capitalism would grind to a halt. A huge percentage of the United State economy is based on consumer spending[4]—in other words, making nature dead and probably going into debt to do it.

Because of humanity's never-ending thirst for more stuff and the shocking volume of resulting waste, all life systems on this Earth are in decline.[5] Because of this overshoot, we will need 27 planet Earth's to sustain current economic trends by the year 2050.[6] What kind of madness is this? 27 planets? Why—so everyone can have custom cell phones and be fifty pounds overweight?

Where is the outrage? Why aren't people flooding the streets protesting for the land that sustains their very lives? For now, in the public discourse, it is "just" trees humans are destroying; at some point, will it be "just" the planet? We all sit silent as corporations decimate the land we need for life without most of us even thinking about it. I believe at some point some insane level of decline will be accepted out loud with a straight face. "Oh don't worry; it's just trees, just the ocean, just the land, just my children, just me..."

Another problem with seeing everything as resources is that it creates pointless competition. Think about it. If we all have to compete, and we do, since we have to pay to live here, we will probably not trust each other, nor will we work together. This can even escalate into wars. Does anyone still believe that the conflicts in the Middle East are over freedom and democracy? Could it have anything to do with oil and other natural resources needed to keep civilization going? I assure you— this is a modern-day trend. None of this was known before agriculture.

An article I read recently was about a study that demonstrated that our pre-agricultural ancestors were not war-like. "Researchers from Abo Academy University

114

in Finland say that violence in early human communities was driven by personal conflicts rather than large-scale battles. They say their findings suggest that war is not an innate part of human nature, but rather a behaviour that we have adopted more recently."[7] Would it be okay for me to infer that "more recently" means when people began to think that they could control the Earth via domestication?

The prevailing myth about "primitive" people from *Leviathan* by Thomas Hobbes is that pre-agricultural life was savage and brutal. Yet, when we observe extant uncivilized tribes today, we see nothing of the sort. However, when I look around at Western Civilization, I see humans enslaving plants, non-human animals, land, rivers, oceans, the sky and vulnerable members of their own species as resources—things to be used, power to be harnessed, and stuff to be had. It is all perpetuated and sustained with violence.

There is no amount of objectifying the Earth that lends itself to a sustainable way of living. Civilization is doomed. 27 planets by 2050 is suicidal. There is no possible justification for doing this or not actively trying to stop it—that is, if you care about your life and your children's lives. We need a global values shift that centers on the fact that the Earth around us is alive and is not ours to use. We can take from it as is needed, but in return, we must protect the land rather than exploiting everything in sight, creating gratuitous possessions and then throwing everything away year after year because a slightly faster model came out.

Think of how you would live if you planned to live where you are forever, and then do it. Would you objectify your land base? I doubt it. Think on this and then tell everyone you know. Enter into relationships with nature instead of seeing yourself as above it, as hierarchies are only maintained and justified with violence. Would you like you or your children to be seen as something existing for someone else's use? If not, do not do it to others. There is no such thing as a resource unless you plan to shield your eyes and ears from the bloodbath that surrounds you or you are okay with someone else objectifying you. This insanity needs to stop. It needs to stop *now*.

This predatory thinking has got to go. It is our only chance of surviving and minimizing our impact on everything around us.

Notes:

1) Responsible Forestry:
   http://worldwildlife.org/industries/responsible-forestry

2) What's the System?
   http://www.realtrees4kids.org/ninetwelve/system.htm

3) Vacant Houses Outnumber Homeless People in U.S.:
   http://www.truthdig.com/eartotheground/item/more_vacant_homes_than_homeless_in_us_20111231

4) Consumer Spending is 60% of the Economy, not 70%:
   http://dmarron.com/2010/06/25/consumer-spending-is-60-of-the-economy-not-70/

5) Data Shows All of Earth's Systems in Rapid Decline:
   http://www.ipsnews.net/2011/07/data-shows-all-of-earths-systems-in-rapid-decline/

6) 27 earths Needed by year 2050:
   http://www.soc.hawaii.edu/mora/Publications/MoraPress1.pdf

7) Primitive human society 'not driven by war':
   http://www.bbc.co.uk/news/science-environment-23340252

19. Science

I want to get to the point of this chapter right away: I am beginning to hate science. This chapter will therefore be slanted in that direction. Behind "green," it is my least favorite civilization buzz word. I hope this does not stop you from reading further, but I understand if you skip ahead now. Questioning science these days seems like questioning religion during the Dark Age.

"Science" is the modern human's catch-all answer. In debates, once a person says "science", it almost does not require an explanation. The debate, from some people's standpoint, is over. Science is the weapon that justifies civilization. Science is good. No science is bad—unthinkable.

What is science? According to Merriam-Webster:

knowledge about or study of the natural world based on facts learned through experiments and observation[1]

That really seems too broad. Let's dig deeper from the same source:

the state of knowing : knowledge as distinguished from ignorance or misunderstanding

That certainly clarifies things. We can combine the two and generally, and I think fairly, we have what most people are talking about when they throw science at me as a defense. So, in essence, it is studying the natural world—figuring it out—because we are apparently ignorant of it.

I believe I can end a science argument with a simple question that I originally heard from author Derrick Jensen: is the world better off because of science? If humans had never made science into the religion that it is today, what would the planet be like?

When I speak with people about this, they point to seemingly wonderful things like indoor plumbing, air conditioning, electricity, and of course, modern medicine. In other words, civilization. That is what science has brought to the world. Of course, in order to achieve these things, humans first had to enslave the resources and their surroundings. Chop down acres of trees to build houses, forcibly extract metals and oil for air conditioning, and practice relentless and cruel vivisection on vulnerable creatures for medical and cosmetic science. Look at the damage science and its pursuits have done, both physically and psychologically.

So again I ask, are we all better off because of science?

Once humans decided that they had the right to manipulate matter, time and space, they began to view the world in a completely different way. Consider, do you think you would ever find uncivilized people holding rhesus monkeys hostage in labs with electrodes hooked up directly to their brains? Would "primitive" people ever conceive of something like a thresher, slaughterhouse equipment, a nuclear bomb, or a boner pill?

When weighing the benefits of science, it is necessary to consider the causes of the problems that science solves. Take, for example, indoor living and all of its modern pairings. Why do we need climate control, refrigeration, metal razors, televisions, and so on?

Climate control: this is pure speculation, but this could be nature's way of telling humans that they should be living somewhere warm. According to the myths of science, human evolution began in the Fertile Crescent, which is what we today call the Middle East. It is very warm there, and believe it or not, it once had land replete with forests, marshes, clean rivers and edible plants.[2] No plumbing, heaters or air vents needed. (Unfortunately, humans decided to domesticate the plants and animals, and we all know how that turned out.)

Do we really need to live in very cold climates in our stately houses with our electric blankets and space heaters? If your answer is "yes," then I understand why you might champion science. Common sense would tell us to live somewhere warm and inviting where the Earth would provide for our needs, not our wants.

Spreading ourselves all over the Earth has only been beneficial for those humans with money and power. It's been really bad for everything else.

So again I ask, are we all better off because of science?

Refrigeration: In a way, it's good that our food doesn't rot immediately. Well, it's good for our modern lives. When acquiring food was a part of our ancestors' everyday activities, this wasn't a problem.

The problem arose when science "freed" some people up to do other things, like make war. Ploughs and domesticating animals allowed fewer people to grow more crops. (Let's put aside how labor intensive agriculture was without modern tools.) Of course, then threshers, tractors and other farm implements were invented, which made food production skyrocket. Since there is now a lot of extra eats, we have to find a way to store them. And, voila, refrigeration was born. Somebody then figured out how to harness oil, which made portable freezers a necessity.

James Howard Kunstle coined the phrase "3000 mile Caesar Salad" to accurately describe food globalization. Basically, companies grow food in faraway places and need cold boxes to do it, hence refrigeration. This is good, right? So it's okay to pillage the planet for all the natural resources that make this happen?

I must disagree. It takes a lot of oil to run this global machine. If you want to see how science has treated the Earth while making refrigeration and worldwide transportation possible, please do an image lookup in a search engine for "Canadian Tar Sands." After witnessing that, realize that at one time, humans ate only what the Earth provided them. They did not grow food; food grew. That is how nature intended it.

Science destroyed that, and now, the Earth is seeing an epidemic of fat people. Science to grow all that food and the science to store it made it scientifically possible for two thirds of adults and nearly a third of children in the United States to be obese.[3]

I repeat: if humans lived according to nature and only ate what the Earth provided, there would be no refrigeration problem for science to solve.

So again I ask, are we all better off because of science?

Metal Razors: Go to any supermarket and look at the magazines. I suspect you may begin to feel fat, ugly and misshapen, plus impoverished and way out of fashion. Maybe, unlike the celebrities, your life is boring because you haven't gotten divorced or arrested this week.

With the rise of civilization came the rise of fashion, and, of course, science (and its friend propaganda) was there to ensure that we had the very latest outfits and accessories. If you're still at the supermarket feeling gross, take a quick trip down the hygiene aisle. There are thousands of different rubs, scrubs, shampoos, soaps, balms, make up, and numerous other chemical cocktails to put on our disgusting skin, as well as razors, shavers, tweezers, scissors, lighted mirrors, straightening irons, curling irons and many other pieces of great science to help us deal with our hideous hair. If you need proof that we should have all these things, go back and look at the magazines again. (If you're feeling depressed about being

121

fat and no longer care, go to the cookie aisle and pick up a box of stuff that contains a long list of ingredients you can't identify. Food science has literally created products to induce chemical reactions in your brain to make you forget your troubles.)

Fashion is global, and it requires a lot of electricity and machines to make it happen. Science to the rescue! It seems every week I hear about a new cream or salve to help me control my unsightly pores and pimples. Scientists come up with these things in labs. Also in these labs you will find little animals who need to find out for us if these new products are dangerous. I think the most well-known instance of this is scientists putting products directly on little white rabbits' eyes in a procedure called the Draize Test.[4] Yup; those scientists smear toxic chemicals into rabbit eyes and leave them there for a while to make sure the products won't hurt humans.

Science has also created a whole host of technology to forcibly sedate and immobilize those rabbits, as well as their friends the rhesus monkeys, dogs, cats and whoever else a scientist can shove a needle into.

All because civilization tells you that you are fat and ugly, and unless you cover yourself in science, your unsightly body might look and smell... natural.

So again I ask, are we all better off because of science?

Television: I stopped watching television years ago, but I cannot go out anywhere without seeing a monitor with some asshole on it trying to sell me something or "entertain" me. People are even inflicting this on themselves with modern cell phones.

Televisions are, of course, made of metals and plastics. Science figured out how to extract all of this from the Earth. Science then figured out a way to put cables all across the land and launch satellites into orbit. Science then found a way to build and maintain a power grid so that we can all watch TV.

Besides the fact that it is all built on violence, television is nothing but a distraction that negatively affects people's health. Science, while investigating problems it created, figured out that sitting down all the time while we watch this crap is really bad for people. "Not moving enough may be as hazardous to your health as smoking, a new study shows." It goes on to say that "6% of heart disease, 7% of type 2 diabetes, and about 10% of colon and breast cancers, are linked to lack of activity." [5]

Don't worry though; laboratory scientists are hard at work coming up with new medical surgeries, diet pills, exercise machines, genetically-modified foods, motivational techniques, and a whole host of other products to help fix this situation. Science will ensure it, and no natural limitations will stop this from happening no matter the cost.

So again I ask, are we all better off because of science?

In one debate, a friend informed me that I am only focusing on the bad that science has done. Science itself is neutral, he said. It's what people do with the knowledge that makes it right or wrong. Sure, science has created weaponized diseases, but it has also improved "our" quality of life. (I'm guessing by "our" he means humans.)

I suppose if I look at it like that, he's correct. Science has made me some fun video games to play, warm socks to wear, and a car that protects me while I go to the store to buy more stuff. If I view the world strictly in an anthropocentric light—meaning I view humankind as most valuable—science seems great. If I break my leg skateboarding, medical since will save me!

Regrettably for me, I don't look at the world like that, which really sucks since I am alive at a time when this "me first" attitude is dominating culture. I think the way I do because I actually asked the land and the other Earthlings how they feel about science. I consulted with them since they have stood by for countless years, stoic, in the face of humans feeling as if they are the masters of the universe. They have seen humans invent spears, swords, guns and bombs with weapons science, which has brought the natural world almost to its knees. The land, the plants, the animals and uncivilized and vulnerable humans have fallen victim to science's rapacious desire to control everything. No matter how much science enslaves and manipulates, it always wants more, and it exists only for that purpose. Any good humans can attribute to science is when science helps clean up the mess it made in the first place.

For the last time, I ask, are we all better off because of science?

Notes:

1) Science Definition:
   http://www.merriam-webster.com/dictionary/science
2) Fertile Crescent:
   http://www.princeton.edu/~achaney/tmve/wiki1
   00k/docs/Fertile_Crescent.html
3) The Surgeon General's Vision for a Healthy and Fit
   Nation:
   http://www.surgeongeneral.gov/initiatives/health
   y-fit-nation/obesityvision2010.pdf
4) The Draize eye test:
   http://www.ncbi.nlm.nih.gov/pubmed/11425356
5) Physical Inactivity May Be as Deadly as Smoking:
   http://www.webmd.com/fitness-exercise/news/20120717/physical-inactivity-may-be-deadly-as-smoking

20. Time

Time is running out. I mean that; it's not just a cliché. The planet is dying and we humans are the cause. Our civilization is going to crash and burn and we will take what's left of the planet at that time with us. Historically speaking, we are an eye blink away from killing ourselves and destroying the very elements we need to live.

How can I get convince people that we are running short on time? Information? Maybe, but I have little hope. We are in the information age and it seems like people only want to focus on celebrities and sports instead of doing research into critically important issues—even if I do the work and then show them. It never fails to amaze me that I can throw some horrible facts at someone and that person is unaffected. For example, during the Super Bowl this past year, which I could not care less about, I commented on a post on a social networking site that scientists estimate that industrial capitalism and human activity are wiping two hundred species per day off of the planet.[1] Forever. They are gone and will never come back. The guy made a joke about it and said it was overblown and alarmist.

Maybe he's right. Maybe that number is high. Let's use a more conservative model. According to the World Wildlife Federation, "[i]f there are:

-100,000,000 different species on Earth
-and the extinction rate is just 0.01%/year
-*at least* 10,000 species go extinct [every] year."[2]

That's huge. Using the archaic food chain model (even though life is actually cyclical), how many links can be chinked and broken before the entire biosystem collapses? Why doesn't anybody care—especially parents? I've always wondered how come people with children aren't radical environmentalists.

126

Maybe if I phrase things differently, I can convey the urgency to the masses. I guess I can try to couch it in other language for our capitalist friends—tell them their way of life is at risk if the destruction doesn't stop. "[UN Biodiversity Chief] Ahmed Djoghlaf says nations risk economic collapse and loss of culture if it [*sic*] does not protect the natural world."3 Of course; talk about something hitting people's wallets and they might change. It is not enough to tell them simply that their activities are the equivalent to them shitting in their own mouths. With the reaction I generally get, I may as well be shouting from the top of a mountain.

I will try again. Still not convinced we are running out of time? Not a capitalist? Good for you. It's still affecting you. How? Well, Monsanto has now decided to use gene silencing technology to kill off more "pests". "Monsanto has applied for regulatory approval of corn that is genetically engineered to use RNAi, as the approach is called for short, to kill the western corn rootworm, one of the costliest of agricultural pests. In another project it is trying to develop a spray that would restore the ability of its Roundup herbicide to kill weeds that have grown impervious to it."4 I'm sure none of this will poison you or your children despite a French study that showed ingesting genetically-modified food causes giant tumors after just a few months.5 (It was retracted later because it was deemed unscientific but not incorrect. 6 I have no proof of this, but I would not bet against some huge corporation threatening the scientists who conducted the study with lawsuits or violence.)

So, to the people who I cannot convince via common sense, capitalism, or protecting themselves or their children, I will try again. (Maybe you are a nihilist philosopher.) In the film *What A Way To Go: Life At The End Of Empire*,7 I heard an amazing and sobering analogy. Paraphrasing, the world is like a building, and we are all blocks. We humans, via our activities, are taking blocks from the foundation of that building and putting them on top. We can do that for a while with little consequence, but eventually it is going to topple the building. I would take it further and say that we are destroying the foundational bricks by driving species extinct. We cannot live for long without them.

Do you understand yet? The rest of the non-human world is running out of time, and if we aren't destroying it outright, we are sullying it with our waste. For example, Fukushima Diiachi is a Japanese nuclear power plant that suffered a tsunami in 2011.[8] It is responsible for dumping tons of radioactive water in the ocean every day, which is undoubtedly mutating any species in the Pacific Ocean who are unfortunate enough to swim in it. (What's going to happen to us when we eat those creatures?) Rain is probably making the situation worse as it is flushing radiation out of the damaged reactors even faster since one of the reactors is sitting on a slope.[9] It is even rumored to now be reaching the West Coast of the United States.[10]

Actually, I take that all back. The truth is, somebody was short-sighted enough to first think up nuclear power and then build radioactive housing units in an area that is susceptible to natural disasters. Then they blame the water! Isn't that amazing conditioning and denial?

That reminds me: I want to get something off my chest. I am sick of people blaming the weather and other natural Earth occurrences for their problems. Here are a few examples of what people generally whine about:

a) Maintenance: Stop complaining because you live near the ocean and wood on your house that came from a forest thousands of miles away is not doing well in those salty conditions, or that metal extracted from somewhere is rusting, or that you have to repaint your boat every year. That is nature degrading what does not belong in that environment. Blaming the weather for this would be like putting a snowman in the desert and complaining because you have to keep adding snow to keep it from melting. It is not the weather; it is the culture of civilization thinking that it can put things all over the Earth without consequence. Entropy is inescapable.

b) Natural disasters: There is no such thing. There are no droughts, floods, earthquakes or tsunamis. This is just what humans call the Earth doing what it has to do and how it affects capitalism. Sorry, but the planet does not care if you built your house near a river or a volcano or close to a fault line. It has no growing season, since agriculture is not natural, and it would be happy to dry up your crops if it determines that an area needs more sun than rain this year. It is common sense to plan for the Earth's needs rather than capitalism's and ridiculous to blame anyone but yourself if you do not.

c) Ruining people's plans: Conditions outside mess up days at the beach, and outdoor events. Actually, that is not correct. The problem here is not the weather. It is that humans have so little free time that if their one scheduled day of fun is cancelled it really is a big deal. But again, don't blame the weather. Blame civilization. Our pre-agriculture ancestors never had to deal with anything like this because they did not have jobs that took up every waking hour of their lives so they could pay bankers and governments to allow everyone else to live on Earth. If it rained at the beach, the ancients probably just shrugged it off and rescheduled to the next day.

d) Tying in with above, weather is unpredictable: Speaking of our ancestors, I have read stories about them knowing when to gather food, when to travel, and when to do various other things based on tidal currents, the wind, and how plants around them grew. This is because they knew the land.[11] They had an actual relationship with it rather than the figurative one most of us have today. Therefore, it is reasonable to assume that generations of people passed down enough knowledge so that life was not so unpredictable. That includes the weather.

e) Cancer from the sun: Evolution originally put dark-skinned people in very sunny climates and loaded up their skin with melanin for a reason. If fare-skinned humans want to live in those same places, they are going to have to deal with the ramifications. Cancer is not an accident; it is the body telling us that something is wrong, or that we are living in an area where we do not belong. Blame civilization, not the condition or the weather.

Seriously, the media focuses on the weather like some kind of fetish. It is hours and hours of excruciating detail and speculation. And, of course, it is never anything important. They never say anything like, "Well, if we hadn't created this ridiculous, unsustainable society, we wouldn't need to worry about snow plows and sun block."

It would also be wonderful if the media, instead of focusing on the weather and pop culture, reported issues relevant to time, such as peak oil, how much time before the Earth can no longer take on any more of our waste, and how much time we have left before Fukushima and other man-made disasters destroy us all.

Do not look for that, though, as civilization acknowledges time only when it affects capitalism. It is easy to control people when you convince them that daily, hourly, and up-to-the-minute schedules are normal. And of course, business meetings, quotas, and hourly pay could not exist without culture perverting time.[12]

I am here to tell you that none of culture's time distortions matter. It is all fiction and busy work used to distract you. Time, to you, should only be relevant to you as to how it relates to the natural world. Civilization will try to tell you that time is only necessary for capitalistic gains. That is not true. The only important aspect of time is to use it to relate to and warn others about just how much damage we have done to the planet and how much longer we have before it all collapses. You need to figure out the most effective way to convince people that our very survival depends on slowing down civilization's activities that will, if given enough time, kill us all.

Every passing hour, more species die. Every passing day, the planet suffers a bit more. Every passing year, we hurtle closer toward the cultural implosion from which we can never recover. Many will suffer unless we make time to stop this.

Hurry now. Time is running out.

Notes:

1) UN Environment Programme: 200 Species Extinct Every Day, Unlike Anything Since Dinosaurs Disappeared 65 Million Years Ago:
http://www.huffingtonpost.com/2010/08/17/un-environment-programme-_n_684562.html
2) How Many Species Are We Losing?
http://wwf.panda.org/about_our_earth/biodiversity/biodiversity/
3) Protect nature for world economic security, warns UN biodiversity chief:
http://www.theguardian.com/environment/2010/aug/16/nature-economic-security
4) Genetic Weapon Against Insects Raises Hope and Fear in Farming:
http://www.nytimes.com/2014/01/28/business/energy-environment/genetic-weapon-against-insects-raises-hope-and-fear-in-farming.html?hp&_r=0
5) A Comparison of the Effects of Three GM Corn Varieties on Mammalian Health:
http://www.ijbs.com/v05p0706.htm
6) Controversial GMO Study By Gilles-Eric Seralini Retracted:
http://www.huffingtonpost.com/2013/12/02/gmo-study-seralini-retracted_n_4357450.html
7) http://www.whatawaytogomovie.com/
8) Timeline: How Japan's Nuclear Crisis Unfolded:
http://www.cnn.com/2011/WORLD/asiapcf/03/15/japan.nuclear.disaster.timeline/index.html
9) Water 6,700 times more radioactive than legal limit spills from Fukushima:
http://www.nbcnews.com/news/world/water-6-700-times-more-radioactive-legal-limit-spills-fukushima-v20797895

10) UC Berkeley Researchers Study Kelp For Possible Radiation From Japan Fukushima Plant:
http://sanfrancisco.cbslocal.com/2014/02/06/kelp-being-studied-at-berkeley-for-possible-radiation-from-fukushima-plant/

11) Primitive Nomadic Peoples:
http://www.bartleby.com/86/16.html

12) Time And Its Discontents by John Zerzan:
http://www.primitivism.com/time.htm

21. Ugly

Throughout my life, I have never really understood fashion. It could have to do with the fact that I can't see properly, but I think the more likely reason is that it just doesn't matter to me. I color coordinate my clothing and generally don't leave the house without pants, but that's as far as it goes. I get clothes used from various places because somehow, my body doesn't corrode when I put them on, and I have saved thousands of dollars doing this.

My hygiene, much like my fashion sense, is nothing special. I buy generic soap and razors. I don't ever wear cologne, and I cut my own hair. All of this saves me tons of cash, but that's not why I do it. Beyond the basics, I just don't care. The media never was able to brainwash me into thinking that what I wear actually matters, not to mention I have no desire to befriend anyone who would judge me using brand names as criteria. (As an aside, it's getting harder and harder to find clothes that don't have somebody's corporate logo splashed across them.)

My clothes, much like our ancestors, are far more utilitarian than fashionable. Because I am big and strong, people always seem to need my help for manual labor tasks. I can't exactly do that in a suit, plus I enjoy the work, so over the years, what I wore mirrored this attitude. I don't own any white clothing, no jewelry, only a few pairs of shoes, and whatever I have is either cheap and disposable or durable enough to endure whatever work I am doing.

As I grew, I always wondered why any of this would matter to anyone. Then, as I got smarter, I realized that this culture hates itself, which is why the fashion and beauty industries even exist. These industries have raked in billions of dollars convincing people that they are fat, that their clothes are not good enough, and that if they just buy this or that product, they will look like whatever celebrity the industry is currently propping up.

If you need any proof of this, go into a supermarket and look at the magazine section. It is page after page of airbrushed models with chiseled abs implying or directly asking why you don't look this way. But, don't worry; we have the solution for you. Buy this chemical cologne and the ladies will swoon. Put this paint on your face and you can hook the man of your dreams. Dye your ugly, ugly hair with our product and you will be the envy of your friends. And don't forget, pull out your charge card for our sweatshop labor clothing and third-world child labor mined diamonds.

Did we forget? Why are you using that out-of-date cell phone, you jerk? What's wrong with you? That car you drive? It sucks. Trust me, bro—you'll never get chicks until we get you into one of *these*. And you're on your way to *that* restaurant again? You know how passé that place is, right?

I hate all of that so much. Fashion, beauty, fitness and advertising agencies have decided what we need. They have decided that the human body is ugly and that it stinks, but it can help us with all the rubbing, scrubbing, shaving, plucking, picking, waxing, tanning, plastic surgery, hair gel, fake nails, pec implants, shampoo, deodorant and all manner of other disgusting shit to help us achieve perfection. Heaven forbid we all look and smell... human.

On top of the fiction fashion has created, it is also contributing to planetary destruction. All of those petroleum-based products come in containers that, of course, are made from raw materials. Since nothing is ever really thrown "away", all of those brightly-colored containers, those old plastic bottles, the used-up cars, and the big piles of old clothing you throw away end up in landfills or in the ocean. Of course, the industry keeps pushing, which means more resource extraction from the Earth, which means more habitat destruction—all in a futile pursuit... for what? What is the end game of all of this? What is the point? Does anybody know?

Am I the only one who is thinking about our marine friends? The ocean is littered with plastic containers, and all of the goo we wash off of ourselves every day ends up as runoff in rivers, which murders whatever is living there, or it runs through barely-regulated waste treatment plants that try to make it so it your child isn't drinking soap and shampoo when you get her a glass from the tap.

The one thing the fashion industry will not tell you is that even if you did end up looking like whoever they say is beautiful, it would not make you happy. Maybe it would in the short-term, but after that, you would still have the problems you already face. Unfortunately, I've seen this many times helping people lose weight. They get down to the "ideal" number but still feel depressed. It is because this culture is sick, and it has invaded their minds. To exert some control, they concentrate on their physical appearance instead of dealing with the problems in their heads where makeup can't reach.

How many celebrities do you know who are unhappy? I won't name any here, but some very famous people have committed suicide, smoked/drank/drugged themselves to death, and still more engage in destructive, bizarre behavior. They have traveled the world, they have all the money and fame and are the fashion industry's standard of beauty, but many are miserable. Seemingly every week, some well-known person proves yet again that money, fame, power and beauty cannot make you or them happy.

Humans evolved to live in the wild in sync with the Earth, and to use whatever the planet provided in a utilitarian manner. They did not own a thousand pairs of shoes and hundreds of outfits; they put on their backs only what they needed to keep warm or cool. They could never have thought of a daily hair and makeup regime—especially with products made up from toxic chemicals. Some were tan, but it was because they were outside where we all belong. Cologne? What's that? They could never have imagined buying a cell phone every year, trading in a perfectly functioning car for a new one, or dealing with a mountain of waste because of this activity. They didn't see themselves as hairy, short, tall, smelly, pale, tan, chic, or whatever. (Obesity was probably unknown pre-agriculture.) They saw themselves as human.

You should too.

22. Vegans & Vegetarians

I love vegans and vegetarians. The many I have met are all wonderful, caring people. They know about some important issues, and they are not afraid to tell people. They care deeply about their fellow Earthlings. They explain to humans not in the know that despite the Cartesian myth[1], animals do feel pain. They are conscious. They are suffering under the current system. They deserve our respect. They must be freed. I really love that—the people, the community, and the activism. It is much needed.

I was a vegetarian and a vegan. The first time, I lapsed. I got burnt out pretty badly and little by little, meat crept back into my diet. The second time, I went full out vegan for a little over a year. It was very satisfying at first, but I ran into several academic dilemmas that I could not resolve. (Please do not hit me with a vegan cream pie.)

Despite the universally good intentions human herbivores have, they are fundamentally misguided. I speak often to vegans and vegetarians (veggies) because I know their language. I have memorized the arguments and I love to see what they have to say. They almost always tell me of the factual, large scale global daily animal brutality and murders—everything from food production to zoos to animal testing. These are vital subjects. Everyone should be talking about them. However, when we get down to solutions, the problems begin.

To most veggies, the way to halt factory farming and all other non-human animal abuses is simply to not eat meat or use products tested on animals. While this seems to make sense, it is, in essence, withdrawal. It is like protesting showers by not taking showers, plus cutting down a lot of trees for remonstration pamphlets that most likely, more often than not, end up in the trash instead of changing lives as they were intended.

Alternatively, the solution I propose to veggies is often met with hostility. That is, we need to dismantle civilization. Civilization has created the conditions necessary to exploit non-humans as they as is being done today. Domesticating plants via agriculture led to domesticating animals, and then herding animals, and then creating—with help from Thomas Hobbes, John Locke, Ayn Rand, and others—the capitalism excuse, which puts a price on life. Capitalism also seeks efficiency, and on paper, factory farming and slaughterhouses look pretty well-organized.

One need only do a search engine image lookup for slaughterhouse equipment for a truly horrific depiction; you will find such items as "chicken slaughterhouse bloodletting line", "meat cutting band saw", and a "skin peeling machine". Civilization created the kind of mental state needed to not only think these up but promote their use.

A question I pose then to veggies is—how is your strategy working? Are companies no longer making murder machines? Is not eating meat stopping the 300-400 cows per hour killed in an average slaughterhouse? [2] Is it slowing climate change from methane that livestock produces? Is eating only plants stopping waste runoff into rivers that destroys marine life?

Further, is it halting the innumerable "bug" (they are animals too) murders from pesticides sprayed onto vegetables? Does it end the industrial corporations' forest clearcutting—excuse me, "forest conversion"—and displacement of both people and animals for palm oil?[3] Does it impede people who are "sport" fishing from barbarically injuring fish and then throwing them back as some sick form of entertainment? Does it cease leather or fur production or end its victims' suffering? Does it stop the unbelievably high number of animals sexually mutilated (we call it "spayed" or "neutered") every year, or the many more killed (we call it "euthanized") in shelters every second of every day? Does it bring to a standstill humans from throwing away and wasting around 90 billion pounds of food per year[4]—I promise you that a lot of it is meat—or make the government, with your tax dollars, stop subsidizing meat production?

If it doesn't, I have serious concerns for your strategy—especially in the long-term. I am pleased that you are concerned about the state of the world, but until we all adopt a lifestyle that does not require cages and factory farms—or until you do something directly like adopting an animal from a kill shelter—you are achieving nothing. You are actually doing worse than nothing since the tactics you engage in make people feel good about doing nothing rather than acting to actually free those who are oppressed. If you want to make companies stop exploiting animals, you have to physically stop it from happening. You'd better get going, since meat production is only increasing year after year.[5]

Any part of a strategy for activism that benefits non-human animals should center on dismantling civilization. At the core of civilization is the right that somehow, humans can and should enslave animals, plants and even each other. With such an attitude comes domination. With domination, there is no equality, and the "might is right" strategy often ends with somebody dead. Naturally, 'weak" humans are not exempt; just look at the numerous times countries with large militaries have invaded foreign countries, killed their citizens and stolen their land and resources. (Please note that not eating meat has never stopped a war. War destroys the environment, which probably hurts a lot of animals.)

After I point out the ineffectiveness of meat abstinence, veggies will next tell me that humans evolved physiologically to eat a less destructive plant-based diet. In other words, maybe we can do less damage to the planet via personal consumption. The problem is—our pre-agricultural ancestors hunted and ate a myriad of plants animals—salmon, buffalo, passenger pigeons, and so on. We evolved eating meat, and when we shifted to more grains and plants, our health suffered tremendously.[5]

What our ancestors certainly did not do was confine thousands of chickens together to live in squalor, use gestation crates to ensure pigs could not move, shock hens in battery cages to make them lay eggs, forcibly impregnate heifers in "rape racks" to keep them producing milk for some other species to consume, or destructively trawl the oceans. They also did not monocrop—that is, they did not engage in agriculture. They, unlike civilized humans, wanted to live on their land forever, so they refrained from destroying it and took only what nature gave. They shared the land with non-humans instead of forcibly keeping others away. That strategy worked for a great many years, and it precluded the "need" for chainsaws, backhoes, slaughterhouses, cattle prods, livestock trucks, hunting rifles and impotent after-the-fact pollution cleanup policies. The attitude was that the Earth owns us, not the other way around.

Unfortunately, it seems that many veggies see industrial civilization as a given. Their strategy for change could be framed like this: "How can we keep capitalism's infrastructure in place while getting everybody to stop eating meat?" However, even if everyone stopped eating meat, the huge, fossil-fuel-based modern human population would still need food, which would require agriculture, which of course kills a host of animals as outlined above (mostly with pesticides and farming equipment). Plant-based diets create hierarchies (animals are more worthy of saving than plants), and I would be willing to bet that at some point, humans would make the connection that if bugs and rodents are expendable if they get in the way of growing food, then why not discard the entire diet and keep right on eating pigs, chickens, cows and fish? Is it really okay if worms, spiders, small birds and mice die in the wake of agriculture so long as slaughterhouses get shut down? To me, murder is murder—either indirectly from clear cutting land in order to grow and harvest crops or directly from factory farming.

Further, the veggie argument that people are starving because livestock eat all the grain grown on the planet is just a capitalist efficiency argument. Switching diets to include more grain would still encourage agriculture, resource exploitation and overpopulation in the West. (Hey; we use *way* more resources than poor countries.) Also, the amount of food available on the planet is not the problem; access to it is. Industrial capitalism needs to be removed in order to feed people, not some combination of restricted personal consumption or large-scale diet change.

It all boils down to this—as cliché as it is: "When the last tree is cut, the last river poisoned, and the last fish dead, we will discover that we can't eat money..."[7] The diet we choose does not matter. We need a values shift. Our attitude needs to mirror what worked for literally millions of years—an uncivilized lifestyle. Civilization must be stopped in its tracks, or innumerable animals will die. No personal diet restriction can get that done.

Notes:

1) Cartesian Theories:
   http://www.iep.utm.edu/anim-eth/#SH1c
2) Inside The Slaughterhouse:
   http://www.pbs.org/wgbh/pages/frontline/shows/meat/slaughter/slaughterhouse.html
3) Environmental & Social Impacts Of Palm Oil:
   http://wwf.panda.org/what_we_do/footprint/agriculture/palm_oil/environmental_impacts/
4) Annual American Food Waste:
   http://www.miamiherald.com/2013/06/17/3456050/americans-throw-away-90-billion.html
5) Meat Production Continues To Rise:
   http://www.worldwatch.org/node/5443
6) Meat Consumption: Evolution & Progress:
   http://www.nature.com/ejcn/journal/v56/n12/full/1601646a.html
7) Quote Origin:
   http://quoteinvestigator.com/2011/10/20/last-tree-cut/

23. What Can I do?

The world can be an overwhelming place. We are born into a violent, coercive death culture that threatens to swallow us whole if we do not fight our way through life. Withdrawing and/or having a defeatist attitude are both easy, but neither are solutions. What follows is a list drawn from my experiences, activism and research. Use what you find useful and discard the rest, but please—do *something*.

1) Do not have children. As well-intentioned as many parents are, modern society exposes children to horrific circumstances. People see violence from birth and are rewarded for participating in that violence. (CEOs of huge, exploitative corporations are generally pretty rich.) In addition, the toll on the Earth is tremendous. I highly recommending watching National Geographic's *The Human Footprint* to witness the vast amount of material it takes to support one human living in Western Civilization. Lastly, to appeal to any capitalists reading this, children are an incredibly bad economic decision. They cost hundreds of thousands of dollars to raise properly. The trade off is not worth it unless you have completely thought it through.

   I mean it. If you do not have significant financial means, do not have children. If you are not obsessed with optimum health for yourself and your baby, do not have children. If you plan on giving birth in a hospital and doping yourself up to avoid the pain, do not have children. If you expect to quickly go back to work and drop your offspring in daycare, do not have children.

144

I beg of you to not do that to another human being who is being brought onto the planet without his or her consent. If you cannot create a loving, supportive healthy environment, spare the child a potential lifetime of misery and pain.

If you already have children, *educate yourself on how to raise them properly.* Also, you or your spouse should get another or a better job, as one of you should be home at all times. I've heard that children can cost upward of a million dollars from birth through college.

If you do not have children, realize that the planet needs your caring and nurturing efforts. Children who are here now are being abused by civilization and modern culture. Having your own will detract from time that you can otherwise spend dealing with some really pressing issues.

2) Live simply. Withdrawing completely is not helping anything, but we can minimize the damage we do on the environment by not owning a thousand things from the dollar store, a giant McMansion, a bunch of cars, and so on. Without all this, you will not need to work as much and can focus on being free or engaging in activism. Alternatively, save your money and spend it on things that will make the world a better place, like giving it to worthy charities.

3) Educate yourself—and I don't mean in public schools. Despite what "green" environmentalists tell you, driving electric cars, changing out your light bulbs and being a vegetarian will not help improve the lives of those currently suffering and dying. Doing things like this is just withdrawing and making yourself feel good but is largely inef-

fective. It would be like protesting showers by not taking one. Find out how you can physically make change and do it.

One of the best ways I have found to do this is doing a search on "anarcho-primitivism" on the Internet and reading everything you can find. Not a reader? There are thousands of free documentaries online. I have watched my fair share and learned many invaluable pieces of information. If you can't watch them or you are often busy, listen to them. (That's what I did for a long time.)

4) Talk to people about what you learn. Today's culture will not ever teach people the things you know if you've made it this far in the book. Tell people. Educate them. Try to relate it to them rather than bludgeoning them over the head with knowledge. If it doesn't work out, keep trying. You'll eventually find people with whom you can relate. If you get discouraged, remember—social revolutions happened long before the Internet existed. It can be done; it just takes a lot of effort and patience.

5) Realize that what you are doing is important. Polar bears, other people, trees, oceans—they are relying on you. Even people in the future will thank people like you for caring. Many individuals become jaded because they do not see immediate results from their activism. Remember: even painfully slow progress is progress.

6) Try to do something meaningful every day. Bringing down civilization is an enormous responsibility that millions of people doing millions of things must undertake. There will never be one act that will send us back to sustainable living (short of nuclear war). Get as many people acting as you can.

7) When you get exhausted from activism, take a break. It's okay. We have incredibly important work to do and it can often be heartbreaking. Realize too that life is also beautiful. Take some time out to hug your spouse, watch a good movie or enjoy life as best you can. Burning yourself out and quitting will not help anyone.

8) Do things that will benefit yourself and other people. We need many people on our side. For example, plant public gardens. Find a way to turn abandoned parking lots into community gardens. Plant rooftop gardens. Learn and teach others what plants are edible. Set up rainwater collection stations. Do whatever you can, for when this culture finally collapses in on itself, people will still need to eat and you can help them.

9) Think small. I remember hearing a story about a Buddhist monk walking along the beach. He was throwing starfish into the ocean who were there because of high tide. Someone asked him, "Why bother? What difference does it make?" The monk picked one up and threw it in the water. "It makes a difference to this one." He threw another. "And that one." Every positive action means something—even if you can't immediately see it.

10) Laugh. It helps. This is heartbreaking work, but the silver lining of the problems being so numerous is that almost anything helps.

This is by no means a comprehensive list. All you need to do is play to your strengths. I enjoy writing, so I wrote this in hopes that people would draw inspiration from it. If you love being around people, talk to as many as you can. If you're an artist, use art to your favor. We need it all. And we need it now.

24. Xenophobia/Immigration

I really hate this subject, but I feel that I need to address it because the political climate in the 21st century has a fetish with it. It is creating so much stress, conflict and covertly accepted racist behaviors.

All you have to do is flip on the radio to any well-fed, loud-mouth, wealthy, right-wing white male to hear about how immigration to the United States is the worst problem ever in the history of problems. That and food stamps, welfare, and anything else that goes against their personal idea of a free and prosperous nation. Those *immigrants*. If we could just kick them all out then George Washington would rise from the grave and salute.

Since those wily Mexicans are right at the border just waiting for immigration guards to fall asleep, let us examine that country, which is apparently the biggest threat to American jobs since Communism. In 2013, the US Census Bureau reported that the United States took in over 280 million dollars worth of Mexican imports.[1] Isn't that interesting? We can take in all of their goods, but the people? "Keep them the hell out of here", thunders the right wing. (I do not think it is coincidental that the pundits promoting or denying this all appear to be making a lot of money.)

The only reason that anyone seems to care about illegal aliens is because of how it might affect capitalism. US workers have to compete with illegals for jobs. I guess if propagating this turn living things into dead things-centered culture is important, then poor people trying to scrape together a living would be a threat that we need to focus on all day, every day.

The problem with immigration is not immigrants. For example, if you refer back to the Property Rights chapter, you will see that we in the US do not own any of this land or any of the goods made with raw materials forcibly stolen from other people in other countries. Property, patriotism, borders and government are all fictional. Those in power created them and then indoctrinated our children to believe and repeat them so the children would identify with the rich capitalists while scapegoating the poor and downtrodden. Focusing on the desperate foreigners distracts people for just long enough for the industrial corporations to continue exploiting the land and the people. Then, the corporations can lower wages and blame the illegals. It works out really well if you are further up the hierarchy. You and your family get rich while the gringos absorb all the heat.

If you take even a cursory glance at US history, you can see that all of the land we are on used to be somebody else's home. The indigenous people and the non-human animals and plants were swiftly and mercilessly crushed under European settlers' boots. At that time, the Europeans were the immigrants. Those illegals caused actual problems, including widespread rape, murder and land exploitation. It is really too bad the American Indians did not have someone to shout over the airwaves. It would have actually mattered then.

There is some sordid irony about white people with European ancestry complaining about immigration, isn't there? I really do wish a tribal leader would emerge today and become prominent. Then maybe we could discuss a real issue rather than supposed threat to jobs and the American way of life (a.k.a. industrial capitalism and the rich getting richer).

Derrick Jensen, in his talks about his book *End Game*, had a good solution to this problem. Yes; let us seal the borders and stop all immigrants from coming into this country so long as we also stop importing the natural resources from those countries too.[2] I wholeheartedly agree.

Speaking of natural resources, immigration is a biased issue in that it does not take anyone except humans into account. For example, humans have moved

149

into all areas of the planet, dominated or destroyed what was there, and then set up industry. Who will speak for the old-growth trees that were in the US long before the European immigrants? Who will vouch for the fish in the rivers that are negatively affected by foreign materials we call "dams"? Where are the talk show hosts criticizing the non-native crops domesticated all around the world and forced to grow in places they do not belong? What about the life those foreign plants choked out in the insane and unsustainable pursuit of agriculture? (We call the native plants "weeds" and murder them.) What about when multi-national corporations immigrate to third world countries and destroy the ecosystems?

I submit that domestication and capitalism are the most insidious immigration offenders ever. Their global reach has poisoned the land, air and sky and has displaced countless numbers of native humans, non-humans and plants. Why will no one speak out against those who practice and promote capitalism as taking away the livelihoods of indigenous people, or destroying their way of life so that they are hopelessly dependent on a system that ruthlessly exploits them? Those voices—including yours—need to be heard. Immigration of capitalism threatens to destroy the planet on which we need to live, and it will continue to do so until we stop it.

If we are going to complain about things moving to where they do not belong, under the ultimate authority of nature, let's get serious.

Notes:

1) Trade in Goods with Mexico:
   http://www.census.gov/foreign-trade/balance/c2010.html
2) Derrick Jensen – Endgame:
   http://www.youtube.com/watch?v=mtuxHVD4Srw

25. Yoga

I've been lifting weights for over 20 years. I started because a bunch of assholes were bullying me. I used to get into a lot of fights because of it, but I figured that if I lifted weights and got big enough, my size would make people leave me alone. My new muscle armor would be a portable, obvious warning to deter anyone looking to initiate violence. For the most part, the armor worked beautifully. I began to feel superior and above-average. (This type of thought is rampant among jocks.)

The problem is that, like with everything else, I took it too far at my health's expense. After getting bored with traditional weights, I began competing in strength sports more than ten years ago and really amped up the training. It got longer, harder, and heavier. I got much stronger. I brought home some trophies and was able to put my anger into something productive.

Now, as I rapidly approach middle age, my body is beginning to break down. My knees are shot, my left shoulder hurts, I wake up in the middle of the night with numb hands, and, after most training sessions, it feels like I've been hit by a truck. I can't stop. That decades-old fear taunts me. "Why would I want to be 'weak' or be average like everyone else?" After decades of pushing weight, this bulk and strength have become me. It's not just part of my identity, but me, the person.

In order to continue hard training, I started to do yoga a few times per week. It hurt pretty badly the first few times. But, I kept it up, and it helped. Well, it helped a little. The war inside of my head that is destroying my body versus the fear of being weak continues.

I tried addressing the mental aspect of my training with meditation. Let me tell you—there is no shortage of people willing to talk to me about yoga or meditation, and especially yoga *and* meditation. It is all the rage. (How many trees have been chopped down for all these new-age books?) So, I tried it. Why not? Despite some serious doubts, I lasted a few weeks. Then, my brain that never shuts up got to me. It keeps asking questions that I need to answer.

It seems the focus of yoga, modern spirituality and meditation are to clear the mind and heal the body. Certain religions posit that if somebody meditates enough, that person will find enlightenment, or true knowledge. I, on the other hand, disagree with all of that. Let me explain.

I doubt that any of my thoughts or anything you are reading are truly original. It is possible I thought something up and wrote it down, but I have been learning and therefore influenced for a very long time. Thousands of books, videos, lectures—you name it; I soak it all in. I actively try to learn every day. Therefore, in all likelihood, somebody, somewhere had "my" ideas first. I just put them together in this particular order. Nevertheless, I now have the information. I am, in a sense, enlightened. I sought the knowledge. This is possible simply because I paid attention to the world around me. I did not get to this state by meditating.

In short, meditation does not work. It is little more than sitting on your ass in a dark room trying to observe your thoughts. This activity, especially if engaged in for hours at a time, is somehow supposed to teach you something or reveal special awareness that cannot be gotten elsewhere.

Believe me when I tell you that you can get data and information you seek easily by reading books, talking to people, watching documentaries, or, better yet, going outside and experiencing life. The only knowledge that is currently a mystery to humans is information that humans have not yet put terms that they can understand. I say this because all of the stuff people have yet to discover or learn already exists except it is in a form that we cannot understand. Nobody has ever really invented anything or thought anything up.

As an aside, humans acting as if they have invented anything are incredibly arrogant. Humans invented airplanes and flight? No; they copied birds. Humans invented ways to go faster? No; they copied cheetahs and other non-humans. Electricity? Nope; lightning was here first. Fire? No; it was already here. Our ancestors manipulated their surroundings for their ends. Really, the list is endless. All humans do is simplify and break things down so that other humans can understand and exploit information.

With that in mind, any "enlightenment" somebody gets from meditating will only help that person cope with life as it is. If you have a terrible job, meditate on it and it will help you accept your circumstances. If your life in general is dreadful, same deal: sit in the dark for hours and hours and you will find your happy place.

Could the problem with meditating now be any more obvious? If not, here it is: get off your ass. Nobody solves any problems by sitting down and quieting the mind, hoping the problems go away. (That goes double for praying.) By being complacent through meditation and "mindfulness," you are hurting yourself. In essence, meditation does nothing. It actually does harm by convincing people that doing nothing is helping them.

Wouldn't it make sense for anyone teaching meditation and enlightenment to tell people about how screwed up the world is? That applies to yogis too.

Here are a couple of conversations that should happen:

Meditation Teacher: "Students, the world is fucked up. I mean *really* fucked up. I mean—I could not possibly tell you in words just how bad things are. We are in an extinction event that threatens our survival. Capitalism is literally murdering the Earth. It is poisoning the air, water and land.

"It is also holding you hostage. You see, that emptiness you're attempting to fill with meditation, as well as your search for self-knowledge, are giving you unrest because we are living in a human-generated prison made of cars, cubicles and houses. We have deprived ourselves of our heritage living in the wild. We belong in nature because we are a part of nature. We are not separate from it. Religion, meditating and consumerism are attempts to fill an empty void that civilization created. We can't use anything from civilization to quiet our minds; we need to go back to the woods where we belong and live in harmony with the Earth. It won't be easy, but we have to do it.

"While I understand that you want other answers, I'm sorry to say, there are none. Civilization is the problem. Living unnaturally is the problem. I assure you that meditation is not the answer. What you are doing by meditating is convincing yourself that you are the problem—that somehow, something is wrong with your mind. There isn't. Still, in a futile effort, you are fighting your instincts that are screaming at you. Let them guide you. Take action. You need to get out of your head and actively do whatever your instincts instruct. They are your true god and your only hope of salvation."

Yoga Master: "Students, the world is fucked up. I mean really fucked up. I mean—I could not possibly tell you in words just how bad things are. We are in an extinction event that threatens our survival. Capitalism is literally murdering the Earth. It is poisoning the air, water and land.

"Your body feels tense and unhealthy for many reasons. First, you most likely sit all day and type at a computer. Our ancient ancestors evolved walking and being active constantly as they foraged for food, ran, jump, played and swam. Recreational exercise was not known to them nor was it necessary. They were active as a way of life. I think they would've seen an obese person as some strange alien.

"Second, you eat a lot of things that I would hardly consider food. Scientists created in laboratories much of what most people put in their bodies. These things are not nutritious. Your body does not recognize them and therefore rejects them. This gives you gas and bloating, modern sicknesses, and it makes you fat. It is only getting worse with genetically-modified organisms. Who knows what kind of negative consequences your body will experience as it attempts to cleanse itself of all this crap?

"Third, you breathe in a lot of terrible chemicals. Think about when you are out walking. If cars pass you, they are spewing carbon monoxide out of their tailpipes right into your and your children's faces. 'Vehicles emit numerous carcinogenic chemicals. Diesel contains benzene, formaldehyde, and 1,3-butadiene—all three are well recognized carcinogens. EPA estimates that vehicle emissions account for as many as half of all cancers attributed to outdoor air pollution.'[1]

"Students, you have been domesticated. Civilization and believing we are separate from nature has ravaged your bodies and made you feel this way. No amount of stretching or posing could possibly fix the damage that industrial capitalism has done to you. I wish I had more answers, but there aren't any. Being alive today is really bad for your health."

Of course, you'll never hear any of that due to cultural conditioning and the dramatic drop in profits this information would have on the new age self-help industries.

Don't be complacent. Your body and mind have evolved over many millennia and are never wrong. Your instincts are there to help you. When they tell you that you are miserable and hurting, find the source of the problems and do something about them. Don't shop, watch TV, eat more trash or sit in dark, sweaty rooms contorting yourself in order to quiet the mind. Get out of your comfort zone and make something happen. You might even save some lives in the process.

Notes:

1) The Harmful Effects of Vehicle Exhaust:
   http://www.ehhi.org/reports/exhaust/summary.shtml

26. Zeitgeist Movement

I felt the need to write this chapter because the Zeitgeist Movement (TZM), which is a leaderless social movement, is gaining popularity. I believe at some point, the ideas that they are putting forward will begin to gain traction. The reason for this is that the rich are getting richer, the poor are getting poorer, the land, air and water are getting more polluted, and, quite frankly, TZM has some great sounding ideas. (That and their free online videos have millions of views.) Let me start from the beginning.

In 2007, a man named Peter Joseph looked around at the world he was living in, became disgusted, and created an art piece in the form of a movie. This movie asked a lot of hard questions about religion, 9/11, world banking, and social institutions.[1] It did not provide answers, but it got a lot of people thinking—including me. I count *Zeitgeist The Movie* as one of the most influential things I have ever seen. It snapped me out of a mental fog that I had been in for decades. Sadly, after eight years of college, I had never even thought to question an institution as fundamental as money—what it is, where it comes from, and so forth. (In addition to enlightening me, *Zeitgeist* proved to me personally that educated does not equal smart.)

In 2008, Peter Joseph, after receiving feedback from fans of the first film who were looking for solutions, created *Zeitgeist: Addendum*. Luckily for me, I began watching these movies years after they came out, so I went straight to *Addendum* for answers. I remember being further educated on the fictional institutions of money and banking; banks actually create money out of thin air in something called Fractional Reserve Lending, which a book called Modern Money Mechanics outlines.[2] It is really startling information, but not entirely outside my realm of thought. I remember when I was a kid and I accidentally burned a dollar bill, I wondered why we couldn't just print another. It's just paper after all, right? The only thing that separates a real dollar bill from a green piece of paper with a cash symbol on it is our faith. We, the citizenry, believe the dollar is worth something—even someone's life. (Since learning this, five dollar chicken Thursdays at the grocery store make me sad. A chicken is only worth five dollars? Worse than that, I can exchange this piece of paper for that chicken?)

Prior to seeing *Zeitgeist*, my previous areas of interest were in researching animal rights through films such as *Meat Your Meat*, which provides a gruesome picture of how the food industry works. Agribusiness is, quite literally, the ritual torture murder of sentient creatures for the value of ink and paper—those dollars again. I went "back to sleep" after learning all that because it is hard information to digest. After awakening again because of *Zeitgeist* and *Addendum* I could no longer ignore what was in front of me. Those films started me down the path that made me realize that all of the institutions around us—nationalism, states, government, banking, laws, social contracts, and even philosophy—are fictional. (Religion I figured out on my own.) They aren't real. They're just stuff people made up and either agreed upon or forced on others. Unfortunately, it's usually the latter; people believe these institutions are factual and they act on those beliefs often with negative consequences. (Example: trading a piece of paper for a chicken's life wrapped in plastic.)

One of the most radical ideas *Zeitgeist: Addendum* and its follow up *Zeitgeist: Moving Forward* showed me is something called a Resource Based Economy (RBE). It excited me. For years, I told everyone about it. It seems like such a great idea. The crux of it is that humans need to work together in order to allocate the Earth's resources in such a way that everyone has what they need to survive, plus creature comforts. In other words, rather than

160

creating vibrating dildos of every color, a million different types of shampoo, useless buildings like prisons and so forth, humans should collectively think strategically about resources. How can the raw materials be put to use in such a way that they will not destroy the environment and fill up landfills? Of course, in order to do this, it is necessary to disregard money. If somebody is starving, you give him a sandwich if one is available, right? Under capitalism, no; throw that sandwich in the trash. In the RBE, there won't be anyone starving because society will plan for food to be available to all. That goes for everything else too; no empty buildings, no trees cut down for porno magazines, and no trash. Ever.

No waste? That sounds ideal. According to the Environmental Protection Agency, "In 2011, Americans generated about 250 million tons of trash"[3], which is almost four and a half pounds per day per person. The RBE promotes fully planned recycling protocols, including creating things with the knowledge that one day they will be discarded, and therefore planning to recycle them rather than burying them in some third-world country. With this mindset, it seems like all humans could live in peace. Why fight over natural resources if we can share them? There will be no need for war if money is no longer in circulation, since war, regrettably, is good business. (Just ask a defense contractor.)

This all seems like common sense. Proponents of RBE wish to plan everything to eliminate scarcity instead of having armies invade countries to take needed resources by force. That *sounds* good, but it isn't. Let me tell you why.

The first problem is in the title: *Resource* Based Economy. What is a resource? It's something that is alive now but probably won't be alive later. A tree, a river, a rock, and a person can all be considered resources. It is a way of looking at the world—basically that nature is there for humans to use. It is a word humans invented to dominate the Earth (including each other). If something is a "resource", it becomes subordinate, which creates justifiable planetary hierarchies instead of recognizing that everything is cyclical and connected. Additionally, how can we enter into a mutually satisfying and beneficial relationship with something that we consider below us?

Think about it. If I see trees and the animals that live in them as resources, I will treat them in a way that benefits me. If I see them as equal individuals who have their own lives to live, I will treat them very differently. The word "resource" simply whitewashes the violence inherent in exploiting vulnerable creatures.

Next is the largest problem by far, which is that the RBE wants more technology, which means more civilization. The RBE is big on automating everything from food production to medical procedures. They say it is to eliminate boring human jobs such as cashiers and janitors, which sounds good until you bring it to its logical conclusion. Automating everything puts us even further away from nature. It also does not reduce the violence associated with collecting the above resources in pursuit of a civilized life. It still makes living things dead in order to uphold a modern standard of living. Let me give you two examples.

An organization called The Venus Projection, which advocates the RBE, put together a frequently asked questions page for energy production. 4 (Already, we're talking about having electricity inside, which has no use other than helping humans live in civilization.) On this page are two ideas that I find disturbing.

Underwater Turbines: somebody actually wrote down that humans should put huge turbines in the Gulf Stream to control sections of it in order to generate hydroelectric power. After I got done laughing, I realized they were serious—especially since it says, "[t]hese slow rotating turbines would have a centrifugal separator and deflectors to prevent harm to marine life." That is bothersome, especially considering the ocean is a planetary system. Try putting a slowly rotating turbine somewhere in your body and see if it will "prevent harm". Dropping giant machines where fish live for any reason will cause damage, not prevent it. So too will getting the resources to create these machines. I looked in all of their FAQs; it doesn't say "leave the ocean and the fish alone unless you plan on improving the habitat by removing trash." Nowhere does the Venus Project advocate living with the fish on their terms and respecting their habitat. To them, marine life is just another resource.

I challenge you to ask a fish what the difference is between a strategically placed machine and the tons and tons of trash already floating in the ocean. I really mean that. Go do it. The fish will give you an honest answer. Alternatively, ask the fish if these machines or this process will benefit marine life in any way. I would be willing to be the fish will tell you that this foolishness will only help humans continue to live destructively in civilization.

If you thought that idea was bad, it somehow gets even worse.

Bering Strait Dam: "A land bridge or tunnel might be constructed across the Bering Strait. The primary function of this span would be to generate electrical power and house facilities for collecting and processing marine products."

Damming anything is not a good idea for anyone except humans (usually white, rich males). According to the World Wildlife Federation:

- "Dams can cause significant environmental damage."
- "Dams disturb natural fluctuations in water flow."
- "Water quality can be degraded."
- "The transports of sediment along the river is disrupted."
- "Reservoirs can emit climate-changing greenhouse gasses."[5]

What all of this means of course if that a lot of marine life will die, as will any humans dependent on the surrounding waters (usually poor and/or indigenous people). I cannot imagine how much damage a 53 mile long[6] dam would do. The idea of constructing one for such trivial human wants is ludicrous.

The rest of the Venus Project reading material has more of the same—more cities (which means more birds will die from windows), more turbines (those poor birds), more resource extraction (again with birds), and more exploitation. More. It's capitalism without money—profit for humans at the expense of the rest of the planet for trifling bullshit.

The bottom line here is that the RBE ideas are old news: humans first, everything else second. It's the same old hierarchies bowing to the altar of technology. It is actually worse than that, as it takes the astoundingly arrogant position that humans have the knowledge and right to responsibly manage the planet.

I hate to break it to anyone who believes in the RBE, but the planet does not need to be managed. It is a self-sustaining and self-correcting system, and it needs the exact opposite of management. It needs "resource" hungry humans to simply get out of the way and live as nature intended. The planet will take care the rest, and it will do a better job than our entire collective could ever imagine.

Notes:

1) The Zeitgeist Movement:
   http://en.wikipedia.org/wiki/The_Zeitgeist_Movement*
2) Modern Money Mechanics:
   http://www.rayservers.com/images/ModernMoneyMechanics.pdf
3) Municipal Solid Waste Generation, Recycling, and Disposal in the United States: Facts and Figures for 2011:
   http://www.epa.gov/osw/nonhaz/municipal/pubs/MSWcharacterization_508_053113_fs.pdf
4) The Venus Project Energy Ideas:
   http://www.thevenusproject.com/technology/energy
5) Dam Problems—Environmental Impacts:
   http://wwf.panda.org/what_we_do/footprint/water/dams_initiative/problems/environmental/
6) Bering Strait Basic Facts:
   http://psc.apl.washington.edu/HLD/Bstrait/bstrait.html#Basics

*As you may have noticed, I do not make a habit of sourcing facts to Wikipedia. However, I made an exception in this case as most of this information is personal to me except the movie dates and a few other small details.

www.ingramcontent.com/pod-product-compliance
Lightning Source LLC
Chambersburg PA
CBHW070856180526
45168CB00005B/1846